Donald Alister Macdonald

Gum Boughs and Wattle Bloom

gathered on Australian hills and plains

Donald Alister Macdonald

Gum Boughs and Wattle Bloom
gathered on Australian hills and plains

ISBN/EAN: 9783337313722

Printed in Europe, USA, Canada, Australia, Japan

Cover: Foto ©berggeist007 / pixelio.de

More available books at **www.hansebooks.com**

GUM BOUGHS

AND

WATTLE BLOOM,

GATHERED ON

Australian Hills and Plains.

BY

DONALD MACDONALD.

CASSELL & COMPANY, LIMITED:
LONDON, PARIS, NEW YORK & MELBOURNE.

CONTENTS.

	PAGE
IN RIVERINA	7
FROM A WESTERN HILL-TOP	36
A MELBOURNE GARDEN	49
GIPPS LAND WOOD NOTES	73
JUVENILE POACHERS	92
SILVER GUMS	113
VILLAGE AND FARM	132
KINGFISHERS AND PIGEONS	158
THE HOME OF THE BLACKFISH	173
A DAY IN THE BUSH	193
SUBURBAN WALKS	208
SOMETHING ABOUT SNAKES	227
ALONG THE SOUTH COAST	241

PREFACE.

These sketches have appeared at intervals in the *Argus* under my initials "D. M.," and in the *Australasian* under the *nom de plume* of "Gnuyang," the latter being in part an expression used in one of the Western native dialects to indicate a gossip. The notes are rather those of an observer than a naturalist.

> "Whence gathered?—The locusts' glad chirrup
>   May furnish a stave;
> The ring of a rowel and stirrup
>   The rush of a wave.
> The chaunt of the marsh frog in rushes,
> That chimes through the pauses and hushes
> Of nightfall, the torrent that gushes,
>   The tempests that rave."

Written in moments of respite from the duties of daily journalism, defects of hasty preparation must be apparent, and the reader will kindly accept at the outset an apology for occasional repetitions. A botanist may consider the title faulty, since the Gum is more correctly a Eucalyptus, and the Wattle an Acacia. For the title and throughout the book I have, however, used popular names, which serve quite as well as scientific terms for identification, and are generally more expressive and appropriate. I may further explain that sketches dealing with kindred subjects have in several instances been grouped under a single heading. Amongst many who have been kind enough to encourage the continuance of this country chatter my thanks are especially due to Mr. DAVID WATTERSTON of the *Australasian* and Mr. JULIAN THOMAS ("The Vagabond"). If the book has any interest it will speak for itself; if not, no amount of prefatory remarks can console the reader.

D. M.

The "Argus" Office,
    Melbourne.

GUM BOUGHS AND WATTLE BLOOM.

In Riverina.

THERE is no part of Australia claiming an individuality more complete than Riverina; and under this title may be included the great plains stretching to north and south from the Murray River. One may wander in many parts of the continent and yet fail to discover anything that, with a greater dearth of material to produce effect, leaves such a lasting picture upon the memory. Some Australian bush scenes, quite idyllic in their wild beauty, will be forgotten when that long day's ride in the Riverina country is still fresh in the memory. And who that has ridden across the Old Man Plain, and wondered when the placid sea of grass would curl into a billow, or the sail of an inland ship—the hawker's white-ridged waggon—break against the blue sky-line, needs any description of it? This waste land without limits, summer-bleached and desolate, the monotonous chaos of these endless plains leave impressions that can never be effaced. What Nature gives she gives in plenty here. The clumps of box-gums clinging together for sympathy; the desert acacias, the weeping myalls, standing singly out on the plain; these are cast lavishly over hundreds of miles. The free open life in this lone land has a charm for many men, and they stick to it as long as they can manage to climb into a saddle. Only those who know the real conditions of a boundary-rider's life can appreciate the realism of one of the finest of Lyndsay Gordon's poems, " The Sick Stockrider." A complete history of the territory would be full of romance, and some very hard realities. Stories of lost travellers dying of thirst are much too common to be worthy of special notice. I used to know an old Scotch shepherd who for years had tramped across these plains with his dogs—a pair of splendid collies that never by any chance got a kind look or word from their crabbed taskmaster. One day death met the patriarch on one of his sun-beaten journeyings, and out in the solitude of One Tree Plain the old shepherd sank down and died. The dogs, unconscious disciples of the soldierly rule " death before desertion," nestled down beside the body and slowly starved to death through the summer days. By-and-by a couple of nomadic shearers, striking away towards the " back blocks," found in their path this little heap of mortality.

Nowhere in Australia will a shower of rain effect a transformation more rapid and complete than in Riverina. One week there is a barren waste stretching league upon league, the next a thick green carpet of succulent herbage. Where not even a withered wisp of kangaroo grass appeared a few days since there are now beautiful plots of rare summer grass. If Riverina bore such a crop everywhere the pastoralists would be rich, and saltbush would be a despised plant by comparison. Along the banks of the watercourses, that not long since were merely dry depressions in the land, wild melons are springing and spreading so rapidly as to give the place the appearance of a garden. The taste for these melons is hard to acquire, no doubt. Ancient shepherds and boundary-riders, who have lived for a quarter of a century on the plains, pretend to like the fruit, but their palates have been ruined by an everlasting diet of mutton and dyspeptic damper, and they have long since forgotten the flavour of a genuine melon. Another fruit of fraudulent type growing on the plains is the quandong. Something in shape and colour like a small crab-apple, it is fair enough to the eye, but in taste thoroughly insipid. On some stations they are enthusiastically made into jam; but this indicates a plentiful lack of other fruits rather than any merit in the quandong. The nuts are more valuable than the fruit, and stockmen sitting round their camp fires at night carve them into all sorts of quaint little ornaments.

These central plains are the home of the great red kangaroo. " Great " to science, but not always so in size; for most of those that range in little mobs along the retired bends of the Edwards, the Murrumbidgee, and the Wakool, are not large. The fur after death loses the beautiful red lustre which has given a name to the species, and is the distinctive colour of the males. A delicate pink flush in the under neck is even more rapid in its decay. It seems to be a dye issuing from the pores of the skin, rather than a fixed tint in the fur capillaries. The does are of a delicate indefinable blue. So much fleeter are they than the males that in summer, when the ground is hard, the veteran kangaroo dog will pick out the brighter-coloured males, although experience tells him that the chances of a hard fight at the finish of the run are thus increased. It is a matter of legs, not of chivalry. Most dogs will chase a half-grown rather than an adult kangaroo, and sometimes when the flying doe throws her "joey" from its pouch the dogs turn upon the little one.

In midsummer the kangaroo are either lolling in the cool avenues of the thick box clumps that spot the plain and stretch in streaks by the river side, or they sit in the myall shadow, moving with it as it shrinks at noon, and extends again when the sun is lowering to the west. Seen thus for the first time at a distance, and through the glare of a December heat, they look like the statues of some strange dead gods. It is so hot here that even these timid creatures will not leave their wretched bit of shade, and only take flight as a last extremity; but the experienced kangaroo dog is wise enough to leave them alone at such a period unless he can come to very close quarters indeed. To one who has coursed hares with a greyhound, or held a straining staghound in leash with a deer in sight, it seems incomprehensible that dogs should watch a kangaroo bound away through the steaming heat, and at the same time conquer a natural impulse to follow.

Riverina plains are cruelly hard and bare in midsummer, and striding over them means blistered feet and broken toes. It is amusing to watch an old dog steal upon a kangaroo that is clinging as long as possible to his favourite shade-spot. When the head of the kangaroo is turned from him the dog advances quickly, but always in a crouching position, like a pointer approaching quail. As the game looks again in his direction the dog seems petrified, and the slightest sign of movement or alarm has the same result. Sometimes they remain thus for a minute—kangaroo and dog motionless as carved stone. The head of the squatted marsupial again turns listlessly away, and again its enemy darts a few paces nearer, until a final rush brings him alongside the quarry almost before it is fairly in motion.

In winter, when the ground is soft after rains, the kangaroo falls while attempting to turn sharply, and it is then that the professional hunter and his pack, whose purpose is scalps and extermination, kill the red kangaroo in hundreds. In the evening you find the hunter riding home, blood-bespattered from head to foot, and with a long string of scalps to sell to the vermin boards or the station owners. Skeletons of dead marsupials cover the plains. Could the spirits of the generations of aboriginal hunters, who lived and died here before squatter and selector, sundowner and shearer came, revisit the plains, they would find fit weapons for ghostly warriors in the long white shank-bones gleaming through the grass—appropriate gnulla-gnullas and boomerangs. In some places are bleached bones piled high beneath the trees. They mark the site of a battue—a sacrifice offered after the old Pagan way to King Merino of the Golden Fleece.

The everyday slaughter of kangaroo with dog and club is not sport. These uneventful massacres are not worth remembering, but a burst on horseback over the plain in the moonlight lives in the memory.

Of winter nights there is a crispness in the air that gives extra zest even to the exhilaration of a ride across the plains. No fear of danger here from low-hanging limbs or rocks and boulders in the path. The grass is thick and soft and green, so that the hoof-beats hardly break the palpable, impressive silence. No night fogs to afflict the lungs; everywhere draughts of purest ozone, which even the weakest may inhale. In summer, when the sun invests the plain, and only the skeleton of the spring verdure clings to the crushed earth, one used to lie prostrate while the blood-red moon climbed sluggishly up the curve of the dead, saffron sky. Then one fancied the dome of the sky oppressively close down, pressing on to one, and the shafts of yellow light were thrown across it solidly, like lamp shadows on the white ceiling of a room. No sea breezes fan this parched land as on summer nights near the coast. But in winter the light is even and alike everywhere. There is life in the night. The horse wants to gallop for his own pleasure. Where the knees press the saddle you can feel the sharp upward roll and forward lunge of the bunched muscle which tells when the steed is galloping high and with his heart in the work. The air is really quite still, yet the sharp breath of motion is humming through one's very hair in merry music that seems to react upon the blood. It is an experience of a winter's night in Central Riverina, and of nowhere else, and even the flying kangaroo, which we race with rather than hunt, is the better for the run. What delicacy in the shadows cast by that metallic moonlight, which has more of silver in it than any other light, though the faintest burnishing of gold gleams close up to every object. The spectre grass is faintly outlined; a silken hair held aloft in the moonlight seems almost to cast its shadow. There are no bluff edges anywhere; the outline is soft, suggestively faint. And in the prospect there is the same want of limit. One might speed for leagues and find the same dim horizon opening up before the same indefinable curves of distance on either hand, the same succession of miles cast behind. A single red star on the plain—the homestead light—is a beacon that beckons rather than warns us away. Such a night ride is the poetry of station life thrown into contrast by many pages of dullest prose.

While the primeval "dog-leg" fence of the Victorian bush, or the latter-day "chock and log," are no impediments in the path of our foresters, the skeleton wire fence of the plains, so fragile in appearance, yet so terribly tough in fibre, has always been a mystery to the red kangaroo. Every day he meets it upon the plain, yet it remains unfamiliar as ever, and almost impenetrable. In flight every other barrier is taken in his bound, but this he rarely crosses. Shirking the leap, he prefers to run down the line of the fence, so that where kangaroo are thick there are marks on either side of many toe-nails. Watch a mob of red kangaroo moving awkwardly across the plain. When moving slowly no animal is more ungainly in its gait, but at full speed all ungainliness disappears. A flying doe bounding away at her best pace is as graceful as a hare or deer in rapid motion. Coming to the wire fence the flock halts, and the wings deflect either way in search of a gap. Those that have become accustomed to one stretch of country know the gates and slip rails just as a hare moves instinctively to a gap in a hedge, or run-hole in a stone wall. Some of the kangaroo feebly thrust their heads between the wires in a vain effort to force a passage. The smaller ones perhaps get through, but to the adults, unless they are much alarmed, a well-strained wire fence is an effective barrier. When a mob is thus cornered in the angle of two wire fences they turn back in the face of the danger, and break past in extending fan formation until, as they speed across the plain, the wings are a half-mile apart. The wire fences are, indeed, responsible for the death of almost as many kangaroo as their chief enemy—the professional hunter. The boundary-rider, cantering along the limit of his run, finds the wires broken, and knows that only the steel tendons of an old man, as thick as the shank-bone itself, could have done it. Sometimes there is a terrible struggle before these iron bonds are burst asunder, and the triumphant veteran shambles away over the grass.

In the anatomy of the kangaroo simplicity of joint is noticeable. The short upper leg-bone works on the shank through the very slenderest of ridge and socket connections, but is wonderfully braced and clamped about with sinews. A joint pliant, but not subject to dislocation. Yet with all this, escape from the toils is the exception, not the rule. Generally the unfortunate animal is found with both hind legs hopelessly twisted in the wires, and the shank-bones broken. Perhaps it has hung thus for days, enduring the concentrated miseries of thirst and hunger, with an agony of pain—reduced to a mere skeleton, yet still alive. As the horseman approaches, the poor animal, worn down by days and nights of struggling, makes yet another feeble effort for that liberty which would be useless even if obtained. The deer-like eyes roll feebly, and the agony so mutely borne touches even its worst enemy. With a pitying, "poor devil," the stockman slips his stirrup, and a blow between the ears ends at once the kangaroo's misery and its life. Sometimes only a white skeleton, picked clean by the crows, marks the spot where the flyer of the plains took her last leap. Before the feeble life was spent these scavengers of the northern rivers had torn the eyes from their sockets and commenced a meal. The eye is their *bonne bouche*, and the carrion birds devour this dainty morsel, whether of merino or marsupial, first.

The fighting kangaroo is one in a hundred. The old man's legs must have failed him in flight before he trusts to their powers in battle. The fiercest fighter I ever saw "stuck up" against a red gum tree in the delta of the Edwards and Wakool. It was during the rains of late winter, when the watercourses were running full, and at some points the two rivers had spread their waters over many miles of plain. The dogs were not inclined to go to quarters with this fine six-foot specimen, and while one of us threatened him with a stick in front, the other crept up behind the tree for a tail-hold. Once get a tail-hold of the largest kangaroo, and he is conquered. He may leap into the air, but a pull brings him back to where he sprang, while at any moment a sharp twitch to one side throws him off his legs. In an instant the dogs rush in, and, keeping clear of the crescents cut in air by his sweeping hind legs, slowly worry the life out of the game. This one warded off the blows with his nervous little fore arms as cleverly as a champion pugilist, and at last, springing clear over the dogs, made a dash for the watercourse. As he plunged into the yellow waters the dogs were once more by his side, and again the "boomer" wheeled and backed against one of the big trees that stud these hollows. While he touched the bottom, and had his scarred fore arms clear, the dogs were swimming. He had them almost at his mercy now, and the usually soft eyes were blazing and the nose twitching in a way that showed the old man was more angry than afraid. As the dogs neared him he bent forward and pushed them viciously beneath the water. There was no flurry, no mistake. Each dog, as he rose, was quietly patted down again —now the brindle, now the fawn, with calm impartiality. The latter had less of the bull-dog blood in him, and soon tired, but the brindle kept up the struggle. With only one dog to deal with the kangaroo would have soon finished the fight. Clasping the brindle in his arms, he held him now beneath the water. The owner of the dog must interfere, or lose his favourite. Rushing in waist deep, he aimed a swinging blow at the old man with a waddy or stunted gum. There were no arms to ward off the blow, but it fell a few inches short; on the crown of the skull it would have killed him instantly, but it struck the kangaroo just above the nose, crushing in the bone, and sprinkling the muddy water with a rush of blood. A spasmodic leap, one last lunge forward at his enemy, and kangaroo and dog went down together. They rose apart out in the current, but neither wanted fighting now. For nearly a quarter of a mile they were whirled along towards the Edwards, then they swung round in an eddy to the farther bank, among floating leaves and rushes, and, side by side, crept ashore; there they lay, within a few feet of each other, both badly injured, both beyond the reach of interference. The dog crept home next day with his side frightfully gashed; and the kangaroo had also disappeared. In that last lunge the old man had struck his enemy just below the ribs, and the long dagger-claw had cut like steel through the flesh; the wound was stitched, and in a week or so the dog, whose sides were seamed with black scars, was winning new laurels, and multiplying his wounds.

Shooting kangaroo with a bullet on the plains requires nice calculation, for all the plain is a dead level, and the monotonous stretch of grass deceives the eye. Perhaps you have scared him from the timber, and when you reach the edge of it there he stands out on the open plain 500 yards away. You get a rest for a careful shot, take a sight just under the shoulder, and then find the lead cutting up the turf 100 yards short of the mark as you fancy, but double that distance out in reality, and going, perhaps, pretty close to the head of the game on the *ricochet*. This miscalculation of distance in Riverina is the experience of the novice who has been accustomed to the downs and broken country of Southern Victoria. Here, as elsewhere, one soon learns to estimate distance correctly, and the man who can bowl over kangaroo with a rifle is likely to be a dangerous sharpshooter in the sterner game of war.

B

Between timber clumps there is generally a clear, narrow vista where the moon's rays fall without interruption, and this is a favourite spot for shooting kangaroo and wallaby on bright nights, as they come up to gain the open, and *en route* pass across this lighted avenue, from a broad, scrubby bend in the river that gave shelter by day to hundreds of them. One night, as we waited, something moved out from the timber into the marginal shadow that was all the blacker by contrast with the illuminated band. "My shot," whispered one of the party, as he sighted the shadow, and waited for it to become a kangaroo in the moonlight. It developed into a man, and as he stepped from beneath the trees his foot trod nearer the brink of the hereafter than it had ever trodden before. How that rifle dropped, and how our hearts throbbed! He who had claimed the shot was remarkably quiet and silent for the rest of the night.

In the flocks of red and blue kangaroo that throng the plains northward of the Murray a white one now and again appears, but rarely reaches maturity. There is a white doe in the Royal Park zoological gardens, with an interesting infant of the same colour, that puts out its head and fore-arms from the pouch in a quaint way, and plucks the grass while mamma feeds. The red old man in the kangaroo paddock is not a bad type of his kind. Imagine a courtly Moslem gentleman bending forward for an impressive salaam, his curved scimitar sticking out behind, and you have a fairly correct outline of a kangaroo in easy motion. The red kangaroo has not sole possession of the central plains. Away in the west, towards the Darling River, the sooty kangaroo, with a dull coat of light brown, a lesser tail, smaller head, and handsome brown eye, varies the monotony of red and blue fur.

My first kangaroo was shot at rather closer range than I wished. I had been blazing away for an hour, in the moonlight, at an occasional boomer, and at what was apparently easy range; the wire cartridges seemed, however, to make no impression. Finally, I went down to the sheep tank, and lying in the shadow of the great mound of earth raised by the excavation, concluded to wait for the game as they came down to drink.

But the animals seemed to be suspicious that night, for they all bounded up to the further corner of the tank, at which range I knew that the cartridges, even if placed with the nicest judgment, would only have served to frighten them. I had almost despaired of getting a shot, when suddenly there came the sound of approaching thumps, and before I could turn round, one of the finest specimens hopped gracefully over the mound, and stopped as though petrified within a few feet of me. It was hard to say which of the two was most astonished at that instant; but the kangaroo certainly was in the next, for, without any attempt at aiming, I raised the muzzle quickly, and judging a line for beneath the shoulder, fired. The poor old man made one nervous, agonised leap high into the air, and coming down almost on top of me, stretched himself in a spasm or two and was dead.

Emu hunting is a test of endurance between horse and bird, and with the former fairly fast and in good condition the bird is almost invariably run down. At full speed the emu drops its tail, throws back its head, and seems to run almost erect, the long slender tail feathers rustling like hay tossed about by the forks of the harvesters, and in a way that alarms a horse new to the work. When tiring, its legs straddle wider and wider apart, so, struggling on to the very last gasp, it falls heart-broken.

The emu is run down for the sake of its skin, but with so inoffensive a bird the sport is a cruel one, and few genuine sportsmen care to gallop a second emu. Emus will eat anything. In captivity, especially, they are a study in this respect. While all that glitters may not be gold, in their eyes it is generally hall-marked as something good to eat. After being limited for a time to a diet of broken bottle and bits of iron, there is nothing with which the average emu so loves to amuse his palate as an eighteen carat watch-chain, with nicely-assorted seal and pendants. His gastric juices are equal to any metallic fragment, and the bird was perhaps the original "snapper-up of unconsidered trifles." The hen bird shows much concern for her downy little ones, and often takes them away to some solitary river bend, so that the male may not kill them.

The wild turkey, or bustard, is perhaps the finest game bird in all Australia. It gives the best sport on the Riverina plains. As it stands statue-like in the open, it is almost impossible while on foot to get within gunshot range; but in a vehicle, or with a horse trained to stand fire, the bird is an easy victim. The man is suspected only when on foot, for if mounted he may come within range, riding round the bird in gradually narrowing circles. In the summer the turkey likes the shade of trees lining the watercourses, and may sometimes be stalked.

During the winter it keeps to the open plains, becomes fat, and is a splendid table bird. Sometimes on a frosty morning—and frosts are very keen in this region—you may see the birds in scores, standing on one leg, erect and motionless, in the centre of a wide open space, as though they had been frozen hard during the night, and were waiting for the sunlight to thaw them again.

The most extraordinary of Riverina birds is the native companion, and those who have been privileged to see perhaps a hundred of these birds in one of their grotesque quadrilles will not soon forget the spectacle. It is a ballet where the fairies are all clad in slate-grey, with just the merest bit of scarlet about the ear-lobes. While other birds are always frivolous and ripe for fun, the native companion has long periods of sanity, when his dignity augments enormously. You see him to perfection only on those plains where whitened turf and blue sky touch all round the horizon and make a region of phantasy of their own. This mirage when first seen in all its beauty is one of the most absorbing of Nature's illusions, and its effects in the strange enlargement of natural objects is weird enough. Through a break in this summer sea of the imagination a stately native companion stalks along the horizon, and, though far away in the distance, seems larger than an emu. A second bird, standing not more than a mile distant, is a mere crane by comparison with the strange giant thus created by the sunlight and the distance. The dancing fit comes upon the native companion suddenly. There is no gradual quickening of the pulse. The bird is either a clown or a stoic, but has no intermediate stage. At one instant you would as soon expect to see this unimpressible long-legged bird dance as to see a kangaroo turn a handspring, but in the next instant it is in the throes of the queerest, maddest series of capers that ever bird attempted. The relapse from gravity is contagious. There may be twenty or fifty companions in the flock, and instantly every bird is dancing as though it were his perpetual occupation. Sometimes they break into groups. Were they gifted with the same imitative faculties as the lyre bird you would feel assured that this quadrille was the result of the birds having watched some settler's outdoor dance. One can fancy the melancholy crow on the myall tree close by, acting as the black-garbed dancing-master, and hear him call, " Down the middle —heads across—wings all round-flap." No command would seem too preposterous for these long-limbed sprites. There are many points of resemblance between the bird gambols and that dying institution, the corroborae of the natives, and, no doubt, many of the " steps " of that wild night-dance of the tribes were borrowed from the "native companions." The birds are determined mischief makers, and if driven from one side of a large newly-sown field, promptly alight at the other end, and continue operations.

The plains of Riverina seem to form a sort of intermediate zone between the tropical forests of Queensland and the colder and more humid timbered tracts along the Victorian sea-board. Thus, birds which are familiar also to the southern colony here show the first approach to the brighter plumage seen in northern forests. Not that rich colours are often noticeable, but the russet-brown and dun-grey tints of the birds of southern woods are more rich and vivid. A few leagues farther north this gloss of feather will change to so many rainbow hues.

The laughing jackass, as known about Melbourne, has no more reputation for beauty than for melody. The browns in his plumage mix in one uninteresting patch of dulness. Along the Edward River, however, the markings are more distinct, and in general appearance "the jackass" is an improved bird. If we met him in the blaze of one of those tropical forests about Cape York, we should find him an exquisite, brilliant in plumage, and harmonising with his surroundings—no longer a foolish laughing jackass, but in very truth a great gaudy kingfisher.

This is no exceptional development, for with smaller birds the same peculiarity is noticeable. Where the plumage is already rich other distinctions are in reserve. Nothing could be more brilliant than the kingfisher of the south as it skims over the surface of a quiet stream, its swiftly vibrating wings leaving two distinct lines upon the water, from which its prevailing colours of azure and deep orange are reflected. Farther north, about Torres Straits, we find him adorned with two long tail-feathers, like a hummingbird. One of the Riverina pigeons, as a first step towards brevet rank and a finer uniform, has put up a little black plume, showing that it stands at least one grade higher than its plain, quaker-like relation of the south. The arrangement is quite seemly; for in the glory of a tropical forest a modestly-attired bird would be a dull blot in the picture, and quite as conspicuous, by reason of his plain clothes, as the Queensland bird would be if turned loose with his fine raiment on these plains.

Another remarkable feature in the bird life of Riverina is the way in which the magpie, crow, and jay have confused their identity. Developing evidently from a common source, the species have been so multiplied that by a complete cycle in evolution they threaten to work back again to the old position. Some of the representatives of different orders seem to have inter-married, and the issue are as mixed in the matter of colour as the human families of the great sheep-stations hereabout. Some are distinct, and very black crows. There is no mistaking them; you find them perched in long rows on the red gum trees that overhang the ana-branches of the rivers, waiting for stock that may perish near the water, or the carcases brought down by the floods. There they sit, the embodiment of woe, and breaking now and again into a croaking chorus. Then there is the genuine magpie, always volatile and happy in spite of his garb of half mourning. There can be no confusion about his two extremes of ebonite black and snowy white. In between these two species come the black and grey magpies as they are popularly known, together with a couple of nondescript birds having some of the peculiarities of each with the identity of none of them in particular. There are altogether too many of the jays in this favoured region, and family relations have become frightfully confused.

Notwithstanding an abundance of food the hawk tribe do not seem to prosper in Riverina, and they have degenerated from the hawk of the southern coast, or the cold forests of Gipps Land. The goshawk, with his beautifully-mottled plumage, is not so handsome as the specimens seen about the Keilor Plains. It would seem that birds of prey thrive better in a cold climate, and reverse the rule as regards colour just mentioned as a characteristic of other birds. They are the destroyers, however, and not the victims, so there is no special reason why they should bring themselves into harmony with the inanimate objects about them. With the nocturnal birds of prey in this locality the same peculiarity is further noticeable. The powerful owl is about the mildest type of his kind one could expect to meet, and there is nothing of the startling apparition about him here. If you can only get the big staring eyes out of sight for an instant, you have a very fair model, both in shape and plumage, of a hen capercailzie or Norwegian grouse in its grey summer coat. Another Riverina bird of the same type, the spotted nightjar, is something of an enigma. It seems to be a sort of cross between the hawk, as far as fierce appearance is concerned, and the owl, as regards nocturnal flitting. The bird has a pair of puny legs, however, out of all proportion to its size and appearance, and a bill that would be considered weak even in a little honey-eater. Altogether, the nightjar is one of the ornithological puzzles of this locality.

What a number of swallows skim about the "billabongs" along the rivers in this semi-tropical region! The warm sunshine, and the equally warm nights, when the air is not affected by the cool sea breezes of the south, seem to suit these birds especially. Yonder is a welcome swallow, with a line of snow-white spots across its barred fan-tail, and here too is the still prettier wood swallow. The nests of these birds indicate their habits. Although both are migratory they ramble according to different methods. The welcome swallow builds a substantial mud nest in the hollow of some decayed box tree, and if used only for a single season this would be labour lost. But the little wanderer will return next summer to its home amongst the myalls, and use the same nest season after season. The wood swallow, on the other hand, has a lack of method in his roaming, and as it may never return to this particular locality it throws together a rude nest of twigs barely sufficient for present requirements.

The cockatoos and parrots of Riverina exhibit a remarkable variety of species, but there are certain traits of form and plumage which link them all together as a common family. Leadbeater's cockatoo, with its plume of crimson and gold, is only a very slight remove from the gillar. The big white cockatoo, with its orange crown and beautiful shading of yellow under the strong white wings, has its variations, and if the pedigrees of the parrots could be traced back they would all merge in a common stock. The black-tailed parakeet with its Hibernian mixture of orange and green, the green leek with hues much more pronounced than the one from which it takes its name, and the pretty little plumed coquette that falls such an easy victim to the bird-catchers in the season, all have their intermediate family connexions. Although the cockatoos seem to live in perfect amity amongst the tree-tops and round about the edges of the swamps, they fight each other with blindest fury when placed together in captivity for the first time.

Amongst Riverina birds rarely seen elsewhere is the painted snipe, a small and very beautiful specimen of the tribe of wading birds. Like the Argus pheasant, all its glory lies in the plumage on its back, which glistens in the sunlight, and without the aid of any very bold colours makes quite a superb display. I have never seen a snipe yet that had not an air of settled melancholy about it, and this one is no exception to the rule.

Here also we find the beautiful white heron, with superb feathers, much more attractive, although a plain pure white, than the plumage of any other bird that frequents the plains. The black and white ibis is held in some esteem, for when the graziers find flocks of these birds coming down towards the south they accept it as an infallible sign of an approaching rainy season.

The lowan and the spotted bower bird—two of the most interesting of Australian birds—are also found in the scrub-lands of Western Riverina, though one must have lived long in solitude and spent many days in the wilds to learn much of their habits. The lowan, or mallee hen, is a shy bird of the pheasant type, and as large as a hen-turkey. It keeps almost constantly under cover of the scrub, and its brown plumage, dabbled with white, shelters it effectually here. The bird may have been the shepherd's near neighbour for months without forming an acquaintance.

The presence of the lowan is indicated generally by its egg-oven in some open space. Instead of hatching, as most other birds do, the lowan, which lays in warm summer weather, forms a mound of sand some eight feet in diameter, and on the summit lays from six to a dozen eggs, which are lightly covered with sand. The heat of the sun acts as an *al fresco* incubator.

As the time draws near for the lowan chicks to break the shell the mother attends the mound constantly, and when the chirping of the little ones is heard beneath she scratches away the layer of sand, and in some slight way helps them into the world. The eggs are moved about in the nest occasionally, so that all may get an equal share of warmth from the sun and sand. They are large in proportion to the size of the bird, almost all yolk, and when freshly laid and exposed to the sunlight, have a flush of dull red, but are afterwards tinged with the sand-stains, and finally present very much in colour and smoothness the general appearance of old ivory. Placed in layers in the mound, they are so turned by the hen that the one earliest laid rests finally on top. A friend of the writer, living on one of the western sheep-stations, took some of the lowan eggs to the homestead, and placed them under a hen. As the shell is exceptionally thin, they hatched rapidly, but like most wild birds, seemed to recognise in their foster-mother an alien, and deserted her at the very earliest opportunity.

The bower bird is entitled to notice, if for nothing else, for its strange architectural tastes. Something of the size of a pigeon, its mottled brown plumage is only relieved by a necklet of rose-pink feathers. The nest of the spotted bower bird is hidden away in the heart of a dense bush, and the egg is perhaps the handsomest in colouring laid by any Australian bird. The shell—smooth like the egg of the lowan—is of pale green beneath, ringed with wavy bands of rich amber. Apart from the nest, however, the bird builds a bower of grass and sticks as a sort of special playground. It is an arched avenue, some three feet in length, with the floor-way paved with matted twigs, and strewn with fragments of bleached bones and freshwater shells. Any shred of glass or metal which arrests the eye or reflects the rays of the sun is a gem in the bower bird's collection, which seems in a sense to parody the art decorations of a modern home. Scraps of this kind lost on the plains are certain to find their way sooner or later to the home of the bower bird, and here in skipping along its avenue, hiding under the thatch, or decorating the walls, the bird spends the gayest hours of its existence. Stockmen are so pleased with its æsthetic planning that they rarely destroy the bower.

In the southern part of Victoria May generally brings with it indications of coming winter, but along the Murray banks the signs are still of midsummer. The white baked clay rings beneath the hoof-strokes of a horse as though it were a bed of ironstone, and on the crust of the sun-burnt earth only stray tufts of whitened grass remain. In amongst the timber the ground is thickly littered with dead leaves, and the white rugged carpet adds no beauty to the plainness of the parched land. There is a monotony in the whiteness here that is not a characteristic of all Australian plains. Southwards towards the coast and the ranges we find always a russet tinge in the fading pastures, and stray vernal tints that defy the fiercest rays of a Christmas sun. The plain also may be veined by green hollows with thickly clustering tufts of dark rushes, under the lea of which hares will crouch for shelter on a windy winter's morning. But if it were not for their lining of stunted gum trees these watercourses that drain the table-lands about the Murray would be as desolate in summer as the plains themselves. Far out on the level a selector's waggon moves along, and a cloud of white dust curling up behind it shows that everywhere the land is bare and cheerless. Now the only bits of colour in the landscape are a few of the imperishable sunflowers or immortelles. Hardly could they have been more appropriately placed. In this earth-oven no sapflower could long exist. These dry yellow gems, which would be quite lost in the glory of one of the heathy mounds about Mordialloc, are as conspicuous on the plains as the finest garden flowers elsewhere. We can imagine the splendid prospect presented here in the days before nibbling sheep killed the majority of the wild flowers. Inside the railway fences some protection has been given, and the view from a train passing across the Rochester Plains in spring is something to be remembered. On either hand there is an endless strip of colour with predominating patches of white, crimson, yellow, and purple, while the plains beyond in vestments of finest green look as though they would presently be as beautiful. In summer every flower is gone, and the grass blades shrivel to white dust.

Were it not for the trees this land would be indeed a desert. The older pine trees on the sand ridges are partly decayed, and the dead brown wood mixes strangely with the living boughs. The young pines are beautifully green. In these arid regions they have a less massive aspect than the imported pine usually has. Contesting with them the possession of the sand mounds is a variety of light-leaved acacia, the Murray willow, just now littering the earth beneath with its lemon-coloured blossoms. Down where the Campaspe pours its slight summer contribution of yellow waters into the purer blue of the Murray some of the red gums are curiously warped and stunted. Nature, or some of her agents, would seem to have "topped" the trees in their youth, so that they might bear thicker foliage. These fine old red gums, with their grotesque curving limbs, give the variety that Australian forests, more than any others, require. Their want of symmetry has kept them alive; for when the pioneer tree-fellers and saw-millers passed down here, leaving behind a train of Gladstonised red gums, these gnarled veterans escaped the axe as being unfit for timber. And the tourist by the river to-day is thankful for that as a small mercy. On some of the skeleton trunks of dead trees hereabout you find the outlines of native canoes still, as sharply carved and well defined as on the day — perhaps a hundred years ago — when the black navigator hacked his shallop from the living tree. The portion of trunk thus exposed withers, while the rest of the tree thrives and thickens, so that year after year the outline of the canoe is more indelibly impressed. At the base of the trunk, where the aborigine could work more freely, the canoe is nicely rounded, and the outlines converge to a sharp point as they go upward. You cannot miss these traces of the olden time; they exist everywhere along the Murray banks. There is a good deal of picturesque variety in the form and colouring of the young gum and wattle shoots that sprout from the earth in hundreds all along the river. The young red gums are pliant and willowy, with leaves thin, long, and hard, and a dull chestnut stem; but the large oval leaves of the peppermint are of a tender green, and the fibres come out faintly when the leaf is seen against the sunlight. Where the shoots strike up from the base of a larger tree they are frosted white. Some of the gum leaves hang in massive clusters; others are thinly and evenly distributed. In one variety of wattle the shoots on the upper side are of a dull ruby colour, lightly sprinkled with a silver bloom, but others, growing close down to the water, are of a universal pale green, while the tender top shoots, in either case, are of a light saffron colour. The tan wattle, planted thickly along some of the southern railway lines, grows naturally here, and the clusters of yellow bloom, although not thickly spread, are, for size, colour, and perfume, the finest of all the wattles.

Occasionally, in a quiet hollow, one comes upon the grey, weather-beaten tent of some Bedouin of the bush, or perhaps the dwelling is merely a few branches thrown up against the prevailing wind in rude imitation of the native mia-mia. You cannot mistake the habitation for the summer camping-place of a tourist. The canvas house of the latter is, as a rule, quite new, and his camp fittings almost luxurious. His plate and cup are nicely enamelled and embossed with his name, but when the real "sundowner" haunts these banks for a season, he is content with a black pannikin, a clasp knife, and a platter "whittled" out of primeval bark. The broad patches of reflected light on the surface of the water play strange tricks with one's eyesight. Away down in a bend of the river one sees a swan floating quietly, and an instant later the stately bird is absorbed and hidden in the shifting, sparkling patch of quicksilver, to be revealed again in a minute floating serenely as before. One is apt to linger for awhile watching the weir-builders at work on the Campaspe. The flakes of sappy red gum fly before adze and chisel, and the planks are bound and clamped everywhere with iron in a way that should defy effectually all the accumulated might that can be brought against it in flood-time from the water-shed away south near the Dividing Range.

All the timber country hereabout is an aviary. There is so little wood and so much plain that the birds cluster by the river, and the trees are laden thickly with their houses. Australian birds love the open and the sunshine. The high, storm-swept mountain top may be the favoured home of the black cockatoo, and the dark, lonesome hollow the congenial resting place of the lyre bird, but our denser forests are, as a rule, deserted by song birds. In the chorus of this region, few notes are more frequent or less musical than the petulant, highly-strung treble of the black and white Grallina. All day long his complaining call breaks in upon the sweeter music of the tree tops. The mud nest is a very durable dwelling, for the little builder has kneaded the clay well, and the winter rains make scarcely an impression upon the smooth outside casing of the cup. Another mud nest built very much in the same way is that of the black jay, occasionally seen amongst the gum trees. In the season it generally carries three white eggs, lightly blotched with black and brown. About the trunks of the trees there are scores of busy little woodpeckers ever searching for food, but the many dancing leaf-shadows hide these tiny birds. Beneath the ragged flakes of bark insects cluster, and, like the oft-dunned debtor, are vexed by the presentation of a little bill. In this case, however, the brown-plumaged bailiff seeks not to enter, but merely raises an alarm by tapping on the outside. In the grooves and hollows of the rough peppermint bark the woodpeckers find much that interests them. A stray stick sent whirling amongst the tree-tops tells you how many birds are living there. One of the jays has a note not unlike the lower call of the laughing jackass. This old friend appears to spend its time here chiefly in crossing and recrossing the river; seemingly a colonist of strong federal inclinations, asserting a claim to vote either in the Murray electorate or in Mandurang. The facility for changing domiciles brings with it occasionally strange complications, sometimes humorous, sometimes tragic. I have known men drowned in a desperate swim for liberty, while the police on the bank were compelled to stand by passively because the fugitive had gained mid-stream. Lighter incidents are happily the more frequent. Not long since the dog tax collector rode down to a snagging party moored on the Victorian side of the river, and demanded fees for a score of dogs owned by the crew. There was no protest—no inclination to shirk the payment; but would the collector come on Saturday when the men were paid? He did come, but only to be received with derisive pantomime from the New South Wales bank of the stream.

At certain periods of the year the Murray is covered with flocks of beautiful wood duck—or more correctly geese, since the bill and the plump shape undoubtedly belong to the goose. Sometimes a bronze-wing pigeon shows amongst the leaves; but a few years ago these handsome birds were as plentiful here as they are to-day amongst the thickets of the Upper Murray. Both the bronze-wing and the Wonga Wonga pigeon are hunted so keenly that in a few years they will have become extinct in Victoria. It is unfortunate, indeed, that to so many the only interest in birds is the pleasure of killing them. The very best proof of the extent of bird life along the Murray is the variety and number of the hawks, ranging in size from the eagle hawk down to the kestrel. On the New South Wales side of the river the eagle hawk is sometimes so great a pest amongst the lambs that the settlers periodically burn him out by climbing close enough to the nest to put a firestick in contact with it. Most of the hawks here build in the deserted nests of other birds—generally a crow or a magpie; but last season on the Murray one of them, a black hawk, actually took possession of an eagle hawk's nest, some five sizes too large. The kestrels build in the hollow of a limb, and on a warm day in summer, when the young birds are being fledged, you find them sitting in a row at the edge of the nest gaining a first acquaintance with the outside world. The eggs of nearly all the birds of prey are distinguishable by their rich red colouring and round shape, as a contrast to the sharp points of the eggs of the spur-wing, or golden plover, both plentiful enough about the plains here. Another Murray bird too lazy to build for itself except when absolutely compelled to do so is the sluggish mopoke. Through its own sparse arrangements of sticks its three white eggs are visible. Ultimately, from lack of architectural skill or industry it turns pirate, and thus unconsciously ensures increased comfort for its young. One of the most peculiar of birds' eggs found about the Murray is that of the locally-termed " cat-bird," the shell of which is veined thickly with dark thin threads as though covered with a spider's web.

From a Western Hill-top.

ONE evening early in spring from a Western hill-top I looked down upon perhaps the fairest picture to be seen in all Australia. Bits of the Snowy country are grander, and some of the southern forests are more sublime, but to my mind there is nothing in the land to equal this. Nature and civilisation, each rich in her own beauty, have clasped hands, and conscious of the power of their association, watch without fear the years flitting by into the treasury of recorded time ; and yet even as Rasselas in his happy valley was discontented, so, possibly, many dwellers here may sigh for a change from their surfeit of rural perfection. If there be any such, let them live for a period on the great white plains of the north—

> "Where with fire and fierce drought on her tresses,
>   Insatiable summer oppresses
>   Sere woodlands and sad wildernesses,
>   And faint flocks and herds."

They will return like that other prodigal — all repentance.

When for the first time one is favoured with a glimpse of the Lake Country he is apt, in the enjoyment of the magnificent prospect, to say, "Surely this is the Australian Eden!" To English ears the "Lake Country" possibly brings reminiscences of Keswick and Grasmere, as well as other scenes that will ever remain associated with that trinity of genius—the Lake Poets. My "Lake Country"—the Victorian land of promise—is made up of those broad western acres on which some of the most fortunate of Australian pioneers have pitched their tents. It is the Camperdown country—the land of extensive freeholds, where the rabbits alone have dared to test the question of ownership with the "first families."

From the summit of Mount Leura the long stretches of grazing country dotted with shining lakes make up a landscape, the beauty of which is not cramped by want of space. Centuries ago this hill was crowned with volcanic flame, but Mount Leura is now a cold and quiet sentinel over the peace and plenty of the west. The tears of adamant that trickled down its scorched and throbbing sides still lie about the base of the mountain, giving just a suggestion of the turmoil that raged here when the aborigines cowered in terror before the manifestations of the fire-spirit. The mountain has lost its terrors, but, instead, the seal of beauty has been set deep upon it. The base is thickly draped with bright green bracken, and the summit tinged a russet red, with sorrel and agrimony. The sides are clothed with fragrant banksias, and on the plains beneath the lightwoods, wisely left when other timber was destroyed, dot the green pastures like beech trees in English meadows. Away in the farther west there is a long line of volcanic hills—the frame of the broad picture on that side. To the south the timber thickens and rises into the Otway range, with its primeval forest and deep calm fern-glades. The storms that sweep across the mountain tops barely stir the fern-fronds, the white satin-wood leaves, or the heavy foliage of the musk trees down in these beautiful dells. The rush of the waves from the Southern Ocean in against the sandstone cliffs comes to one faintly. Out towards the margin of the forest the sounds of life creep in :—

> "Hark ! the bells on the distant cattle
> Waft across the range,
> Through the golden-tinted wattle,
> Music low and strange."

From this mountain summit nearly a dozen great lakes are in sight, their waters glistening in the light as the sun sinks. Within a radius of twenty miles there are thirty lakes, ranging from broad Corangamite with its erratic belt 100 miles long, down to a tiny salt-pool that had a name only in the dead dialect of a dead tribe. But the euphonious native titles that always have a special significance are well preserved in the West. The pity is that they are not more universal. There are flowers about the crown of this mountain, altogether distinct from those flecking the plain beneath. I am especially pleased with a tiny mountain orchid like the opened bill of a bird, the throat of dark crimson, deepening down almost to purple, and the tongue a bright yellow, burnished like a buttercup. The broad green plains stretching away on every side are indeed a fair picture. They have been furrowed with a ploughshare only, and no roaring cannon shot has ever burst the glebe. The song the rain sings here is, "I come to wash away no stain upon your wasted lea." May that hymn of peace never alter. And yet, if the need ever comes, Australians will fight as valiantly and fall as worthily as ever the old-world soldiers did in defending home. Should Australia have her hour of trial, it is out here in the West that her stoutest defenders will, I fancy, be found. Those who first mooted country rifle clubs, whether aware of it or not, gauged the national sentiment. There is a patriotism here that cannot, one thinks, live quite so well in cities—a patriotism born of Nature's grandeur, let us say. It was here that Lyndsay Gordon found the inspiration for his poems. He caught the spirit of the place, and, voiced by his genius, it has found a *refrain* in every hamlet and cattle camp of the West. Even the white iron roofs of the township, nestling amongst the gardens at the foot of the mount, make a prettier rural picture than one ever imagined could grow out of such material. As I sit on the mountain top a pair of rose-plumed cockatoos are wheeling in short circles overhead, shrieking petulantly. It is their protest against intrusion. In the hollow limb of one of the skeleton gum trees on the inner side of the mountain there is an interesting family, the cause of all this concern. On one of the banksias a young magpie that has just achieved its first flight is curiously perched, with its head pointing stolidly skyward. It takes a momentary interest in the surroundings only when the parent birds bring a choice morsel. In all the trees about scores of little yellow-tailed, satin-breasted wrens, each intent with its own home mission, make a twittering chorus. What cunning little homes they weave in the pendent branches. On the top of the little round ball there is a cup-like nest never occupied. It is built as a ruse to deceive the predatory shrike, or the native cat, who, finding an apparently empty nest on top, trouble themselves no further, and overlook the little round entrance covered with a piece of hanging fringe that leads to the snug retreat inside.

I have passed many winter nights at the fireside, listening to stories of Colonial life in this region. What perils and privations were borne by the pioneer settlers who came over forests and plains to lay the nation's corner-stone out here in the fair West! Like Gellibrand—who, confident in his own judgment, handed over his share of food and water to his companions before starting out on that long journey the end of which is still a mystery—each one took the risk of fatal failure. Sometimes, in the heat of noonday, and when water was scarce, they quenched their thirst with the acid cones from the sheoak trees. Women, too, had to face the hardships of pioneer life. One of them, with her great-grandchildren growing up around her, to-day tells stories of those early days—when Melbourne was a hamlet and Geelong a couple of tents. Sometimes, while the bullock-teams were away to the sea-port for fresh loading, the women were left for weeks alone—aliens in the midst of savages, and at their mercy. These were women fit to be the mothers of a new nation. For a time the whites lived on sufferance among the blacks; by-and-by the position was reversed. The blacks, it is said, have fallen before civilisation. The degraded remnants of savagery who pick up a living about bush townships are surely our handiwork? They have been civilised with rum. Whatever knowledge we lack of other aboriginal customs, with their burial rites we should be familiar. We are, indeed, a civilising race. When we came here the aborigines covered these wide plains in thousands. Where are they to-day? We have civilised them—they are dead. Their marriage laws were as strict as ours, their morality stricter, their lives healthier, and their sports purer, but we have reformed all this. As a unit amongst these reformers I cannot say that I am particularly proud of the work, but modesty is one of our national traits. The reason we know so little about these aborigines is, that instead of studying we shot them. Old colonists say that some reformers of the old days were rather less ceremonious in shooting an aborigine than a wild dog. Indeed, the latter incident was the more rare—the dingoes had not the same confidence in our good intentions. Buckley, the wild white man, was never so badly disposed towards those of his own colour as when he heard of fresh outrages on the blacks, and news of that kind was seldom scarce. He knew the black people as no one else did, and was their advocate and defender to the last. Some of us Australians to-day know—from those whose word we cannot doubt—that, viewed in the light of their own customs and traditions, the blacks are not naturally the thieving, deceitful, drunken vagabonds of popular fancy. When mothers did not hesitate to give their infants in charge to the native women to carry away through the bush amongst the trees and the wild flowers, and keep them all day long, they must have had no qualms about their being trustworthy. And they were more tender with the little ones, and cared for them more faithfully, than the modern nursegirl inspired by thoughts of her policeman. One can almost believe that these people in their religion blended the most revered of Christian truths with the principles of advanced thought—the orthodox doctrine of an eternal heaven above the clouds with the Darwinian theory of descent from the lower animals—two latter-day extremes thus meeting in the old-time philosophy of a nation of savages. When the white men, by way of sport or to preserve their pastures, destroyed the kangaroos and emus that from time immemorial had been the native hunters' best game in the West, it was not a startling act of retaliation when the black man killed a stray sheep at times. Yet in such cases the old Jewish ordinance of a life for a life was rigorously enforced. Once upon a time, long ago, a native was caught red-handed at sheep killing, and taken to the station so that he might be handed over to the authorities. There had been a wholesale case of poisoning on a run close by, or even this formality would not have been observed. Some " dampers " were baked specially for the local tribe; but the act of Christian charity was marred through the cook using arsenic instead of salt for seasoning. Nasty rumours had got about, and the authorities scores of miles away talked of an inquiry. Instead therefore of putting a bullet through this wretched sheep-killing savage without ceremony, they fastened him to a dead tree trunk with a bullock-chain. One day he knocked the staple from the log, and gathering up the chain got away into the bush; but dogs were laid on to his track and he was run down and brought to bay on a log in the middle of the river with the ponderous bullock-chain still hanging on his arm. When the station people reached the bank the poor hunted vagrant philosophically wound the chain about his neck and drowned himself. The men took some trouble to recover the body, not because the black Spartan deserved at least a grave, but because bullock-chains were valuable. Had this savage been partly civilised he would have stood his trial and proved an *alibi.*

The power of imitation amongst the blacks is so pronounced as to be almost a special natural gift. I have an old sketch-book, in which an aboriginal at my request attempted some pen-and-ink etchings. His figures of both men and animals— and more especially men of his own race—are very correct. They are drawn in solid, with no attempt at shading, but in some fifty figures of native hunters and fighting men the features are all unquestionably aboriginal, while the characteristics of white men are happily hit off. In the sketches representing his own people as they once were, the artist works from memory, and it is interesting to note the distinguishing tribal marks. The Lachlan blacks have a small plume with a long shank fixed to the hair above the forehead, and not unlike the small tuft worn by some of the English hussar regiments. On their bodies there are different symbols painted, some of them having an old-world significance. The rule and square of freemasonry, the trident allotted to the sea-god Neptune, and the crescent, against which the Crusaders once hurled their power, are a few of them. The Goulburn blacks have an anklet of twigs tied just above the feet; the foliage points upward; but the Western tribes, who in a corroborae wear the same ornament, have the leaves pendent. "Buckeen" blacks, from Queensland, are wrapped in opossum-skin robes. Another sketch shows a group of Murray blacks holding a corroborae. The striking point here is that the artist's fellow tribesmen are altogether a finer lot of fellows physically than the others. It is more than probable that this trait is merely the outcropping of the artist's chauvinism. Like the French artist who painted the day of judgment, he has made his own people the best looking among the crowd. Should any one think that a savage has no perception of humour or sense of the ridiculous, a glance at two of the sketches would convince him to the contrary. One of them shows a black fellow chasing a Chinaman, the latter appearing fully alive to the peril of the position. For the patient, toiling

Mongolian the blacks have a serene contempt. In another sketch there is a group of white men of the up-country dandy type. They have reached an exceedingly after-dinner stage of elevation, and the attitude of each is that of the swaggering rustic "'Arry." A swan's nest is so drawn that you have a full view of the inside of the nest filled with eggs. From an aboriginal point of view the inside of the nest is the only interesting part of it. Iguana and emu hunters are shown, the latter either behind a tree, with a spear poised, or approaching the bird under shelter of a bush-screen. A native woman carrying her infant in an opossum robe fastened to the shoulder is another subject, and the outlines of the head and neck of a white and black woman respectively have the true characteristic of the two races. This black had had no tuition whatever in the use of pen or pencil, and his sketches were simply the result of natural aptitude for imitation.

Could any of us have occupied this flint seat on the mountain top half a century ago we might have witnessed one of those great tribal gatherings so often held in the plain below. Let us imagine that to-day we are looking through that long avenue of years as well as from the mountain. The wild spectacle before us will be of this kind.

Tall columns of smoke have shot up from the swamps by day, and at night signal fires have blazed from dead tree trunks. The message stick, covered with barbaric signs, has been sent amongst the tribes as the fiery cross was of old through the Caledonian clans, summoning them to meet together. And this summons, like the red beacon that called the Gael to repel some fresh invader, was never disregarded. As the shadow of the mountain projects further into the plain the tribes come in to the trysting place, until a thousand souls are gathered together. Camp-fires twinkle in the darkness like the lights of a city. The tribes are to hold a grand corroborae. The chiefs have daubed their faces with red clay; as an insignia of rank the tail of a dingo is fastened in the hair, and swings about the shoulders. In front they have a dark swan's-wing plume, in contradistinction to the white feathers worn by their subjects. Shining white kangaroo teeth, taken from the big foresters that were once so plentiful about the forest margin, gleam in their dark matted hair, as conspicuously as diamonds about the neck of a ball-room belle. The women wear necklaces of porcupine quills. The ordinary warriors and hunting men have a strip of red stringy bark bound about their temples, and a hundred long kangaroo teeth strung about the neck clash together as they move. A great central fire has been built up, and round about this the dancers assemble after sunset, all dressed for revelry. On their dark skins, broad white lines have been drawn with anatomical accuracy; and as they move from the shadows into the light it seems as though so many human skeletons, bleached white under sun and rain, were preparing for a Walpurgis-dance out upon the green plain. Through the pierced cartilage of the nostrils the dancers have thrust fragments of a white reed to add to the effect of the grim pantomime. Fastened to their waist-band is a tuft of the long rustling tail feathers of the emu. Then the metallic ring of the music-sticks is heard through the clear air, and the corroborae begins. A dancer rushes from the outer darkness into the circle of firelight, and stands for an instant as if petrified. Every nerve and muscle and limb vibrates, and as the swish and rattle of twigs, emu feathers, and kangaroo teeth reaches a climax, the dancer falls back into the gloom. Then a pair, and next several dancers spring out from the shadows, until a whole tribe is represented in the illumined space. Thus the dancing goes on, each tribe taking its turn. In approved royal fashion, each chief has his jester, a sort of automatic humorist, with a special capacity for mimicry, doubtless. The jester's feelings at the time may be of a morbid or mournful cast, but he must be funny for the night.

The black court jester finishes his parody of the dancing of a "native companion" or the movements of a kangaroo, and then gives the audience one of his best jokes, specially reserved for the finish, and his jokes have one merit—they are his own. A score of lighted boomerangs are thrown into the air, and for a minute they shoot about like great fireflies in the darkness. In mimic warfare, two bands of warriors hurl toy boomerangs and blunt spears at each other; but the weapons are turned aside by the deftly-handled wooden shields or gnulla gnullas. Next, a band of picked wrestlers meet in the open to contend—for honour only. The victor carries off no opossum-skin rug or stone axe as the reward of his skill.

Although strength of arm and limb are the saving tribal "virtues," the bravest warrior and most skilful huntsman yet ranks beneath the seer who told the names of the stars, and read from the broad scroll above the story of a coming season of storm or sunshine. As in the classic sports a wreath of laurel was the visible emblem of victory, so here the best hunter of game takes a single feather only as a symbol of success. The modern "black fellow" reciting a parody of one of his native songs in the hope of a shilling at the finish is only the degraded *simulacrum* of the old-time savage.

As the night deepens the wild music dies away, each tribe retires to its own limits, and the revelry is over. To-morrow a grand hunt will perhaps be organised, disputes that have arisen since the last great meeting will be settled, wrong-doers will meet their challengers in open combat, and he who has taken a life otherwise than in fair fight must defend his own. Then each tribe will retire to its own hunting-ground, until summoned to another great inter-colonial conference.

Such were the ceremonies—as antique, let us say, as some of our most respectable Aryan folk rites—that have died out with the annihilation of an ancient people in half a century. The sable warriors have, according to a tenet of their creed, passed out along the spirit path to a lone island, where, holding communion only with the seals and sea birds, they wait until the day breaks and the shadows depart.

A Melbourne Garden.

> "And the spring arose on the garden fair,
> Like the spirit of love felt everywhere;
> And each flower and herb on earth's dark breast
> Rose from the dreams of its wintry rest."

SHELLEY's poetic reference to the former home of his sensitive plant applies very happily to this garden in which you and I in common with every Victorian have a joint interest. A garden of the best kind, with lawns and upland, groves and ferneries, lakes and rockeries. Upon the lawns the sward is green and close shaven, and beneath ivy-clothed trees are rustic seats "for talking age and whispering lovers" made. The ferneries might have been picked from one of those dark silent gullies in the heart of the Otway Range, and the ponds are peopled with a feathered community as comfortable as cosmopolitan. Regal white swans float in company with dusky awkward musk ducks and diminutive teal, the contrast being as marked as between the tall white trumpet lilies and the little yellow marsh buttons that fleck the margin of the water. From the clefts in the rockeries spring strange varieties of the skeleton cacti that give an air of sterility and decay to their arid native eminences. In the flower-beds the last triumph of modern floriculture, which largely affects variegated foliage, grows in company with some antique plant, the subject of traditions even in the days of Linnæus.

From the highest point of the garden the view is

charming. Across the pool at the bottom of the slope and through a break in the trees one catches the tiles and gables of Richmond, a grey stone church standing out in relief on the summit of the rise, and all the architectural defects of the suburb lost in the distance. A few steps in the opposite direction, and between the trees in the domain you look down upon the beach at St. Kilda. In the peculiar light the horizon seems as near as the coast line, and rises directly above it, so that the appearance is conveyed of an immense tidal wave, even and massive, without foam or motion, just about to topple over upon the dark dots that represent loungers on the pier and esplanade. On the left a line of pendulous willows, fringing the river bank, make up a wall of tender green. On the upland some tame doves are cooing in a grove, and round about their haunt azaleas in perfect bloom—a glorious show in the sunshine—make up a homely idyll in sound and colour. Jasmine, yellow and white, clambers along the western wall, the paler flowers (less plentiful than the others) starring the background like tiny snowflakes on a carpet of gold and green. A long, trailing spray of ivy has fallen away from the wall into close companionship with a stunted holly bush, and the association is apt to remind one of the holly and ivy girl, dear to Irish hearts as having given a theme to their peasant poet. Here, amongst a little colony of medicinal plants, are names which appeal to the memory like some revered, half-forgotten theme in music — meadow rue, wood sorrel, summer savory, sweet fennel, friar's cowl, wild basil, and wake robin—all the old names so thoroughly, unmistakably English

—the remnants of a homely botanical lore which, with its quaint legends and antique superstitions, is passing away. With these ingredients the fairies of old worked their witchery. Down by the lake-side a supple twining spray of native clematis, bearing light lemon-coloured blossoms, its tendrils clasped about the bole of a gum tree, reminds one of bush gullies. This climbing plant, with its hot, pungent wood and almost perpetual bloom, is an important feature in the Christmas decorations of the country. In one corner of the lawn we have a picture in miniature. It is a single English daisy—a bright speck of gold encircled by the purest white, a little gem placed in a broad setting of unbroken green. The esteem in which the tiny flower is held is something more than a sentiment, for the practical gardener, when mowing and weeding the lawns, is careful that no harm shall come to his daisy.

In the flower-beds there is occasional evidence of wild caprice on the part of a gardening expert. Pink petunias are disfigured with blotches of white, and other flowers the originals of which were, perhaps, a plain crimson or purple, are now a curious compound of even more than two garden hues. One can admire these hybrids as proofs of human industry or as evidence in support of the theory of a natural diffusion of species, but very often the crossing of kinds has been so carried out as to be almost an offence against good taste and beauty. Addison sedately observes: "Variations in flowers are like variations in music—often beautiful as such, but almost always inferior to the theme on which they are founded—the original air." A lover of old airs will probably be an admirer of old flowers also. Their beauty is mixed with so many tender associations that they "speak to the heart like the voice of a long-forgotten friend." A bright marigold or the perfume of a wall-flower will bring many buried incidents thronging upon the memory. Down at Laverton last spring I walked through one of those thoroughly old-time gardens, where all was beautiful but nothing modern, and the company of these homely favourites pleased me more than the very finest of modern exotics grown in the heated air of a conservatory. Sturdy plants of that sort are a type of the British pioneer—rugged, honest, and independent, while the pampered flowers that follow in their train may represent the genteel lounger in the haunts of fashion.

The poets' flowers have ever been the common flowers. Of the sundews and pitcher plants that prey upon bright-winged insects, and tempt the butterfly flitting past in the sunlight to destruction, the poet cannot sing. And if only for the sake of the many rhymes we love, we are slow to renounce the old floral faith. Tennyson confesses his ignorance of the new botanical creed, for in his apostrophe to a flower in a crannied wall he says—

"If I could understand what you are, root and all, and all in all,
I should know what God and man is."

Fortunate, perhaps, has it been—for our children at all events—that our poets knew the flowers more as wood fairies than as competitors in the struggle for life.

In Tennyson's references to the flower world there

is much of rich fancy and imagination, but more of the poet's homely observation. He is not one of those who believe that only the unattainable and the unseen are beautiful. Too many of the poets have sought in mythological fields for the beauties that Nature, even in their own woodlands, had supplied. Their flowers are praised not so much for a distinctive beauty of their own as for being beautiful by comparison. They are, perhaps, "fair as the fabulous asphodel." If these poets have an ideal in fruits it may be the golden apple, guarded by the Hesperides; but Tennyson tells us that:—

> " Lo, sweeten'd with the summer light,
> The full-juiced apple waxing over mellow,
> Drops in a silent autumn night."

Thomson, throughout the "Season," as might be expected, speaks much of flowers, but his "feet" are often better than his botany. He has whole pages where anemones, auriculas, and ranunculus cluster about each other. Reading his poems is like walking down a broad garden path, with flower-beds stretching away upon either hand. Shelley's garden is a more artistic and delicate conception, for in imagery he is a millionaire among poets. Most of Shelley's flowers come to us as creations. Tennyson's are more of a remembrance. With Tennyson a verse, or sometimes a single line, is of itself a complete little vignette from an English landscape. His woodland etchings are at once poems and pictures.

With "In Memoriam" he struck a deeper note. The man whose memory inspired this finest of latter-day psalms must, like the poet himself, have been much above the average of modern men. The poem is a garden filled with flowers as beautiful as the thoughts of which they are an allegory. A garden for all the seasons. In the autumn scenes there are leaves, sapless and yellow, fluttering down to the earth like a fading hope or a dying cause, and the bright chestnuts patter to the ground "through leaves that redden to the fall." Poplar and primrose, witch elms and woodbine, rose and reed, haw and hazel, all cluster in this ample garden, which is at once garden and " God's acre."

Many of the poets are accustomed to associate the lily and the rose, but in Tennyson's poetry we more frequently find the rose with the jasmine as a companion. So it is in several instances, and notably in " The Princess." The anemone is another of the poet's favourite flowers. In one of those queer verses in dialect, "The Northern Farmer," the old man tells the story of the murdered keeper being found on his face, " doon in the woild enemies."

One of the poet's favourite flower fancies is to speak of them as standing out from the surroundings as a light in the darkness. In " The Two Voices " there is the following example:—

> " Not less the bees would range her cells,
> The furzy prickle fire the dells,
> The foxglove cluster dappled bells."

The idea is frequently repeated. In the poem where Œnone, the daughter of the river god, appeals so despairingly to her mountain mother, Ida, there is a versification of the mythological story of the three goddesses Herè, Aphrodite, and Athena coming to the bower of Paris, to ask the shepherd which shall claim the golden apple inscribed "to the fairest." In this incident Tennyson deals with the flowers of fancy as well as of fact—

> "At their feet the crocus brake like fire,
> Violet, amaracus, and asphodel,
> Lotos and lilies ; and a wind arose,
> And overhead the wandering ivy and vine,
> This way and that, in many a wild festoon
> Run riot, garlanding the gnarled boughs
> With brush and berry and flower through and through."

Again, in the "Dream of Fair Women,"

> "Growths of jasmine turned
> Their humid arms festooning tree to tree,
> And at the root through lush green grasses burned
> The red anemone."

The quotation, while illustrating the idea already mentioned, is a fair specimen of Tennyson's woodland fragments.

The "Idylls of the King," rich as they are in chivalric fancies, have few flower scenes. There is the tumult of the tournament, the clang of steel armour, and the ring of jewelled swords, but the pomp is all of man's creation. The flowers are generally typified in the fair women of these mediæval times. The only lily is "the lily maid of Astolat," the only May blossom that on Lynette's brow, companion to the apple blossom of her cheek. Enid was

> ". . . Like a blossom vernal white,
> That lightly breaks a faded flower sheath."

Gawain sang of a rose that was "wondrous fair"—but the rose was a woman.

The subject of "Aylmer's Field" is a war of roses, but in this case the flowers typify two factions in the nation battling for the right to lead—

> "When the red rose was redder than itself,
> And York's white rose as red as Lancaster's."

The poem is not all a *mêlée* of blood-splashed roses. There is a bit of rural felicity in a land of hops and "poppy-mingled corn," while the lines that follow convey a picture of rustic peace.

To realise more clearly the character and effect of Tennyson's flower poems and rural verses, let us imagine a man tired of the world wishing for some picturesque haunt, such as the happy valley of Rasselas without the discontent. In the works of the standard poets he may find the material to build up an idyllic home to compare with the finest saintly picture of the Christian heaven. But it is as unreal as beautiful. In Tennyson's verse, however, he may find bits of rustic mosaic for a material paradise—a paradise of which others would say, "I have seen just such a scrap of woodland, just such a bit of meadow, just such an association of familiar flowers, somewhere in the world."

I have wandered away into other floral paths, fancying that those who can appreciate the charms of this Melbourne garden will also enjoy the Laureate's garden scenes. Coming back to my fern gullies, here we perceive that the ferns themselves are not arranged with the same exactness as in the gardens of those sententious experts whose first principles are a rule and compass, who do ever mix up floriculture with geometry. Their gardens remind one of a prim old maid—everything regular and neat, but nothing natural or pretty. In my Melbourne garden ferns are strewn about lavishly as with Nature's hand, and such confusion is good. The collection is very complete, ranging from our own giant tree fern down past the parasitical elk-horn to the pretty little maidenhair which we find growing in native luxuriance on the nooks and crannies of a precipice. In the heart of one of the fern glades I have a glimpse of an English thrush with a snail in his bill hopping towards the darker shadows. Down by the edge of the lake another thrush has taken refuge under the spreading lower branches of a small cedar, his mottled breast flecked with the sunshine that peeps through the spines. His mate, clinging to a spray of ivy on a gum tree close at hand, is busy amongst the insects. A fourth, forgetting his native modesty, has hopped out upon the open lawn. Seemingly the emigrants have prospered quite as readily here as in English hazel thickets; but the gardeners tell another story. After many years of the experiment there were only a dozen thrushes in the garden last nesting season, and this year there are not more than a score. On the hill in the old garden they built low down, and were at the mercy of every mischievous schoolboy. One day the nests would be full of speckled eggs, and the next they were empty and deserted. Now the birds build in the taller trees about the fern-glades, but here also they have many enemies. In the drains and rockeries there are native cats that make the trees their hunting-ground by night, and destroy both eggs and young birds. Shells of broken eggs beneath a tree are a sign that the native cats have been abroad, and although many are taken in snares set for rabbits it is difficult to reduce their numbers. I more than suspect that the pair of laughing jackasses perched upon a tree near the crest of the hill do much unsuspected damage amongst their feathered mates. As they fly about they are followed by the smaller birds piping shrilly and raising a general alarm—the very surest sign that they look upon " the jackass" as a common enemy.

At the edge of the lake I am agreeably surprised to find well-grown saltbush flourishing more luxuriantly than upon the plains of Riverina. There the saltbush country is little more than a thicket of dry denuded brambles, for as soon as the sweet young shoots appear they are greedily nibbled away. What would a few thousand acres of these masses of silvery foliage be worth up near the tropics now that the suns are parching the pastures?

Amongst the trees the conifers make perhaps the finest show in the gardens. Some of the cedars are frosted with silver, as you may have seen Australian trees before sunrise on a morning of early spring, but never in the glare of noonday. Upon the uplands the trees of many countries are grouped—as an illustration of universal good fellowship. The Irish juniper mingles its foliage with Crimean and Corsican pines, and the American golden yew, with a tinge of bright yellow lighting up its branches, compares strikingly with the dark massive aspect of the English yew tree. An arbutus, which in May and June will be covered with pearly blossoms, is partly hidden in the shade of one of those peculiar ti-trees whose white trunks have a skeleton appearance when seen in the moonlight. A Victorian will probably have made his first acquaintance with them while rushing through the Sydney suburbs in the southern express. Here, too, in a semi-tropical garden, are the characteristics of arctic vegetation. The branches, instead of springing outward and upward from the tree trunks, all droop in the spiritless fashion that gives such an aspect of settled melancholy to northern forests. For generations they have been borne down by the snows, and the tree has finally adapted itself to its surroundings by inclining its boughs downward, so that the feathery snowflakes, instead of accumulating, fall upon the white waste beneath. If it were possible to transplant one of our mightiest tropical trees to the frigid zones, what a wreck would remain after one of these long, dark northern winters. Amongst colonial conifers the Norfolk Island pine is perhaps the most striking. It is unfortunate that the name should have been associated with such dark epochs of colonial history as are described in Marcus Clarke's powerful work, "His Natural Life." The blot has been wiped away, but the name of Norfolk Island seems destined to be indelibly associated with cruelty and crime. Some of the trees are at least as antique as the Scriptures. Here is a scion of the pre-historic locust tree which, according to some, furnished food for the Baptist in the wilderness. On the same bank there is a "Judas" tree.

On a suitable mound there is a beautiful group made up of the native heath and its affinities, all in full bloom. The cultured gardener pauses in his crusade against the weeds, and with the easy volubility of a professor of the dead languages hurls a series of botanical names at one of the handsomest heaths in the collection. Having no grudge against the plant, however, I forget what he told me. All the flower-beds are, in a sense, disfigured by the white labels on which the pedigrees of plants are recorded, but in this garden the educational has, I know, to be combined with the picturesque. Sometimes, after much Latin, the English name appears in parentheses thus :—" Viola Cœrulea (Common Violet)." In noticing the grouping of the trees here it is easy to understand that landscape gardening is no simple art. The altitude of the trees and the blending of the foliage for broad general effects have both to be considered. The tallest tree is placed in the centre of the bed, and the height declines as the clump extends, until the tiny shrub at the extreme edge mingles its foliage with the grass-blades on the lawn. I have always been surprised to find so few bees amongst the flowers here. In the country such a garden would be invaded by hordes of busy little workers, drawing the staple wealth of bee colonies from the sweet white clover that lights up the lawn, the broad sunflowers, or the pretty blue forget-me-nots. The bees are quite æsthetic in their love for the sunflower, only their regard is based more on material than sentimental grounds. They seem, also, to catch the delicate odour of clover-fields on the breeze, and will travel miles for the wealth of these pastures. The dwellers on the heights around sadly neglect their opportunities in the way of bee-keeping. They need only provide a suitable apiary, and here in the gardens all through the summer the bees will find the food that suits them. There can be no charge for agistment, no impounding for trespass, and the flowers would benefit by the visits of their best friends amongst the order of fertilising insects. The glass conservatories on the hill are not open to-day, for even this spring weather is too keen for the sickly-constitutioned exotics. Through the steamy tears that trickle down the inner side of the glass doors one has an indistinct view of the tropical plants, but, compared with the flower-beds outside, there is very little colour in the collection.

This garden, besides being typically Australian, is something more. People of all nations find something in it to admire. If the English thrush could speak as clearly as he sings we might imagine him telling this story.

"Two years ago I believed honestly that the whole of the habitable globe was contained in a little corner of the county of Hampshire. Then I fell into the hands of the bird-catchers, and crossed more leagues of sea than I had ever even dreamed of when hopping about in the copses at home. I shall be a year-old colonist this spring, and qualified to write, as even younger colonists do, both the past and future history of the land. I have but a simple object in piping to-day—merely to describe the past winter experiences of an English thrush in a city garden. It is none too soon to write either, for the night of winter is passing away, and amongst the flowers there is the dawn of spring. The balm of opening buds is in the fresh air, and the green of the springing leaves lights up the foliage as though it were the season of flowers. The box brush is putting out its tiny pea-green bells, and I hear them tinkling fairy tunes when the breeze comes brushing through the trees. Soon the oaks will be 'at home' again. Even now you may see the illustration of the poet's fancy of a thousand emeralds breaking from ruby buds. The rows of elms that border the broad walks are still asleep. The lattice is closed, the blinds are down, the doors yet barred; but after a few such sunny days as this even the sluggard elms will be astir. A young friend of mine, a honey-eater, whom you may recognise some day in her green gown and white earrings, has just come from the country. She tells me that the wattles and white sassafras are blooming by every creek, and that the yellow winter waters are becoming clear and yet more clear with each succeeding spring day. Some of our vain friends in the bush will almost live now in the branches that overhang the stream, so that they may see the brighter hues come out in their summer vestments, and like Narcissus admire their beauty as mirrored in the pool.

"One misses some of the things that give a charm to the English spring. Poor goldenbill, my rival in the hedgerows, is no longer a trouble, and even the Indian minahs, with their quaint, sociable ways, are a poor substitute for an English blackbird. When bird-fanciers pass by I hear sometimes of a caged starling, and remember Sterne's story that the schoolmaster read to the boys one afternoon as they passed down the lane. My heart bleeds when I think of the same melancholy plaint from my poor caged countryman in the city.

"The winter here was so different from those English winters over the sea, when nature, wrapped closely in her white mantle, goes to sleep for a season. There was a darker green in the trees, a deeper shadow in the glades, a keener wind sporting down the pathways, but otherwise little change. Everywhere there were bits of colour lighting up the garden plots. Late in the season came the white flakes of laurustina, with their nutty fragrance. Berries—scarlet, ruby, white, pink, and golden—sparkled in the shrubs upon which no snowflakes had ever fallen. The pink berries, I hear the gardener say, are Tasmanian. This is not a land of nuts. One looks in vain for hazels and chestnuts, but the brown nuts of the pittosporum grew in plenty about the paths. They have given way now to very fragrant pale flowers. In the early winter there was lavender and heliotrope, and another shrub with buff, trumpet-shaped, waxen bowers that made a brave show in spite of the winter winds. Then came the Eugenia with its slender, plum-coloured drupes. The coprosma, radiant in a dress of gold and green, made a glorious centre-piece for the flower-beds, and no blossom could have enhanced its beauty. The wormwood, silvered nearly white, reminded me with other things of home.

"Every morning the clerks in Government offices came trooping down the pathways, and although I sang them my sweetest songs they never seemed to hear. It was very disheartening. They talked always of some grim confraternity of tyrants—members of the Legislature, I suspect. Day after day they came and went along the asphalt, not even noticing that the moss grew on one side only, and that the banks with a northern aspect were bare and cheerless. They stop sometimes, now to look at the long willow tresses or the forget-me-nots sparkling in the flower beds. Once a grey-headed old man, moving slowly along the path, as he heard me sing, lifted his eyes and exclaimed, 'Surely it must be a thrush. How it reminds me of the stile and that old wheatfield down in Sussex. Poor Minnie!' Then he went away, mumbling to himself—

> "'Many a summer the grass has grown green,
> Withered and faded our faces between.'

He passed out into the noise and smoke of the city, and I never saw him again, but his words set me thinking. I remembered English fields just as beautiful as his, and thought how in the autumn I should miss the corn-cockles in the wheat, the poppies, the red campion, the wake robin, and all the pretty corn-flowers. Sometimes one almost wishes to forget these things, as the volatile sparrow tumbling about in the dust does. How daring the sparrows are, and how very much at home! They almost brush against the children playing beneath the trees. Their confidence sometimes verges on presumption. But even they have one trace of home affection left, and they show it in their inclination to seek the ivy that clambers about the church yonder for a resting-place at night.

"Sometimes ladies pass by, carrying upon their bonnets beautiful birds embowered in a perfect paradise of flowers. But how solemn and silent the birds seem. Their staring eyes never blink in the sunlight or sparkle in the shadow. They are not as happy as they should be. I am not generally superstitious, but things happen so strangely sometimes that they set even a thrush to wondering. One night I was asleep on a pine-bough, when a cricket on the turf just beneath me burst out with its loud chirrup, and I woke. On the same bough, within a few feet of me, was a native cat, and but for that cricket's night song I should not have told this story.

"My friends say that at first wild cats were a great trouble to them, but we have found out that they cannot climb the smooth bark of the straight young gum trees, and for the future we shall build there in safety. How happy we should be in the gardens but for these native cats. They are as bad as the weasels and stoats of tne hedgerows at home. There are few owls here to come in the night breaking our rest with their rustling wings, but the laughing jackass is almost as merciless. In England our gaitered friend the gamekeeper would have shot all these tyrants mercilessly, but here no one attempts to do justice upon them.

"You will think it must have been lonesome sometimes in the winter, and so it was. But most of us here—and especially the trees—are aliens, and we tell each other stories of our own homes. The Indian hawthorn there—with haws that are more like olives than the purple bunches I remember in the hedge—is very eloquent. The weeping pine speaks sometimes of venerable Buddhist temples, and sometimes of dark jungles in Southern India. One of its robust cousins boasts of how his ancestors first raised their heads away up in the solitudes of the Himalayas. That melancholy tree in the corner, beautiful as it is to-day in the sunlight, has cruel stories to tell of Siberian prisoners. The Japanese cedar, that shades out to silvery brown, has legends old as the world itself, and yet less familiar to Christian ears than the wonderful tales of that other Cedar of Lebanon. That hemlock on the left complains that he misses the ring of the snow-shoes and the tinkle of the sledge bells in the Canadian moonlight. The ungrateful tree is not even thankful that his branches are no longer torn down with the weight of a thousand snowflakes. The Irish yew down in the corner is a charming talker, but never satisfied. One cold morning I remarked how the frosts were browning the prairie grass, but he insisted that the correct title was Tipperary grass.

"Latterly I have had more company. Some brown doves have come from the domain on the southern bank of the river to make a home in this garden. You perhaps may only feel their presence in the soft coo-coo that gurgles from out of the thickest pine; but I know their haunts, and have called to see them more than once. We are more often heard than seen; but if you come on a windy day, when the trees are tossed about, or after a shower, you will find us on the grassy lawns. Come while the spring is yet young, and our best notes are fresh from a winter's rest, and we will give you such a chorus as was never heard before out of England."

It may at first seem a rash thing to assert that city people have the same chance of studying the habits of birds as the residents in those country districts where the town sportsman goes to shoot away his holiday, and not infrequently his arm or leg. "Dead game" has been the subject of many a picture, although the artists may never in their lives have flushed a flock of wild duck from a pool, or been startled by the hum of quail in their country home. The big gun has brought all sorts and conditions of game birds to the city market, and there, no doubt, most of the game studies have been made. But it is not in the market-place alone, nor yet cooped up in the cages at the Zoo, that Melbourne birds are to be seen. The use of a field-glass from one of the city towers will probably discover such flocks of game as in most country districts would bring every rusty fowling-piece from its cobwebbed corner, and send every second resident off on a wild hunt after the birds. At certain periods of the year the rushy water-pools near the boat-sheds in the domain, and the lake in the Botanical Gardens, literally swarm with water birds, and all the year round some are to be found there. They show a wonderful intelligence in dating their arrival here conjointly with the opening of the shooting season, for they know that on these weedy, stagnant pools—the "billabongs" of northern rivers —no one dare use either fowling-piece or swivel gun. To-day, perhaps, the Melbourne birds are spread from the Murray to the coast, but in a few weeks we shall have them back again, and teal, black duck, and spoon-bill will ornament the city lakes as well as city tables. The pools near Prince's Bridge are more suggestive of the natural haunts of the birds. No doubt they were the home of many water birds before the city was founded. Red-legged porphyry coots are either splashing about amongst the reeds at the lake margin, as they do in those greater waterpools of Gipps Land, or feeding upon the green grass of the open, as tame almost as barn-yard fowls. Out near the St. Kilda railway station I have seen landrails and spur-winged plover amongst the reeds and swamps near the line, and no doubt both birds come here occasionally, but they are not often seen. There are few of the big coots at present, but they generally come in numbers towards the close of summer, and thus observe the fashionable and certainly rational practice of spending their summer in the country and their winter in town. The little black coot, the water-hen of Gipps Land, where it is seen in vast flocks, is another city visitor. Flocks of forty or fifty birds occasionally settle down in the night and stay, perhaps, for months. In the Botanic Gardens one family remains in possession all the year round, and their nests, not unlike the warm home of the wild duck, have occasionally been found in the island thicket. These are the birds that venture amongst the feet of the children as they play about the lake. If they hatch at all very few of the young ones survive, for the white-tailed water-rat has his home in these pools also, and destroys both eggs and young birds. Standing here sometimes in the moonlight, one hears a splash in the water, and just has a glimpse of the water-pest before he sinks beneath the surface again. To-day a matronly wild duck is out with her little downy brood, and in less than a week she is childless, for the whole family have fallen victims to the water-rat. Even the swans, splendid as is their defence of home, come to loathe this imp of the water and darkness, whose destructive efforts are not checked by a single repulse, or even half a score of failures. As he burrows from beneath the water, like the platypus, his home is seldom found, and the chances of extermination are lessened.

The history of the wild ducks who visit here would no doubt be interesting. Those presented to the director of the gardens have been pinioned, and enjoy a sort of limited imprisonment for life. Amongst them are some of the richly-plumaged mountain duck found in every part of Australia, but more naturally associated with the woodlands and waters of Lake George, in New South Wales. There are some curious hybrids between the Japanese black duck and the English mallard, showing how bird species are multiplied throughout the world. They are handsomely coloured, and a distinct ornament on the water. The successful cross here indicates, I suppose, that away back in the family histories of the black duck and the mallard there is some connecting link. Like the little water-hens, these half-domesticated birds are always on the look-out for crumbs and biscuits. While the English and Japanese ducks thus live in perfect sympathy, neither will associate with the Australian ducks. The antipathy is mutual too.

The history of the swans in the garden lake is a repetition in results of the struggles of the black and white races all the world over. Some twenty years ago the first of the snowy-plumaged English birds were turned out here, and for a time their reign was undisputed.

One night a black swan sailing over caught a glimpse of water which had, perhaps, been his ancestral home half a century ago, and dropped down amongst the islands. Everything was changed, but not for the worse. He liked the place, and stayed. Other swans came in the same way, and so two rival groups of black and white were formed. But the pond was not large enough for both to live in amity, and soon there were royal battles between the male birds. The whites, although the more powerful in appearance, were beaten all along the line, and thus history showed us another Hayti, a black conquest on a small scale.

For a time the British bird scarcely dared to show his plumes upon the water; but as he was the greater attraction of the two, besides being more rare and valuable, the Great Powers came to his rescue. The one way to make the immigrant happy again was the old scriptural plan of destroying his enemies off the face of the earth.

It was noticed that in the battles which took place every two or three weeks one of the black swans always acted as leader. When the first of these fighting chieftains was killed it was hoped that the trouble was over, but another took his place, for every man in the black army carried, apparently, a marshal's baton in his knapsack. By-and-by, however, only their widows remained, and once again Britain ruled the waves. As an Australian, my sympathy is with the colonial bird. It shocks one's sense of fair play to find good hard fighting so rewarded. To-day there is but one black male on the lake, and upon him the spirit of philosophic opportunism seems to have fallen. He declines to take any part in the game of "Heads you win, tails I lose," and lives at peace with all birds. Indeed, some visitors to the gardens, concerned for the national honour, have suggested the addition of a white tie to his black garb, in order that this spiritles bird might be named "the parson." The subsequent history of the first family of imported white swans is a sad one. One morning the four white swans were found floating dead in a little arm of the lake. Some ruffian had thrown poisoned food to them as he passed, for analysis showed arsenic in the stomachs of all. During thirteen years the white swans produced only two clutches of cygnets.

The lake is also associated with a dark page in the history of another group of water birds. Once the big black cormorants and the smaller white-breasted divers came here in hundreds to feed upon the golden carp and eels. All day they splashed, dived, and feasted, and at night perched in the ti-tree scrub, at the lower end of the lake. In the dusk the thicket looked like some strange, dark orchard, where the trees bore at the same time large white flowers and fruit. Either the lake with its fish had to be given up to the cormorants, or their raid stopped.

One evening a party of crack shots drew a cordon about the place, and at a signal the slaughter commenced. Dead and dying birds rained upon the water until eighty had thus fallen. The story of the massacre seems to have been handed down through successive generations, for ever since they have shunned the place. Tempted by the goldfish, half sheltered in the reeds, one occasionally drops down, but never remains over night. Down in the pools near the boat-sheds they muster in scores, and remain there through the year, except in the nesting season, when they go out to build with the pelican upon the rocky islands of Bass's Strait. Other fisher birds have had an unfortunate experience here.

One morning a big grey pelican was found in possession, no doubt congratulating himself upon having at last discovered the fisherman's Eden. When shot, sixteen large goldfish were found in his pouch.

Outside the aquatic birds, all the more conspicuous songsters here are foreigners. A notable exception, however, is the reed warbler, a modest, sweet-voiced little native, who has found a suitable asylum amongst the reeds that line the lesser pools. In the dedication to his " Bush Ballads and Galloping Rhymes " Lyndsay Gordon speaks of a land

> " Where bright blossoms are scentless,
> And songless bright birds."

But all Australian bright birds are not songless, and some of her homely ones have a chant of surpassing sweetness. This little brown warbler of the lake has a thrush-like trill and variety in his note that is very charming. Now and again one gets a rare glimpse of an English blackbird. There are only two on the south side of the Yarra, and, unfortunately, both are male birds. The speckled grey doves, that came in the first instance probably from Ceylon, have found a home, in which they prosper and multiply. These birds are spreading over all the city gardens. A lot of English ring-doves were turned out a few years ago, but seemed to have died away altogether. Canaries were equally unfortunate. Their colour seemed to have excited the deadly enmity of other birds. The Indian minah is as much at home and almost as presumptuous as the sparrow. He came with the reputation of an insectivorous bird, but has, as is alleged, developed a strong partiality for fruit. Last year a bird supposed to be a nightingale was seen in one of the shrubberies. It is an interesting fact that during the last few years many native birds that had almost abandoned the place have returned in numbers to their old haunts by the river side, thus indicating the success of the measures taken for their protection. A keen naturalist would probably find at least fifty species of Australian birds within the limits of the Melbourne Gardens.

Gipps Land Wood Notes.

A TRIP through the Gipps Land lake country commences, by rights, at the point where the Glengarry and Thompson rivers lose their identity in the broader bosom of the Latrobe. Anglers from the city have spent many pleasant hours upon the banks of the latter, for perch, ignoring the traditions of their kind all the world over, rise at an artificial fly like trout or salmon in British waters. The lowlands of Glencoe—a most contradictory designation—on the one side, and the long flats of the Heart estate on the other, although both famed grazing grounds, are not rich in picturesque elements. Down towards the river mouth there are long miles of moving reeds, that look like stretches of giant corn over-ripened in the summer sun and still waiting for the reapers. From the shelter of these reeds, in the old days, the wild blacks dashed out to spear the cattle introduced with enormous trouble by the white pioneers of Gipps Land. But here as elsewhere superior human intelligence devised a remedy and a retaliation. Look-out trees from which one of the settlers, by means of a primitive system of flag signalling, guided his armed companions to where the Warrigals were crouching in the reed beds are still pointed out upon the uplands; and in these collisions between the two races there was always the same ending. On the banks of the Latrobe the grazier is yet supreme; but away across the lakes, upon the picturesque flats of the Mitchell, hops and maize fight for possession. In the fields there are tall hedges of hawthorn crowned with bunches of crimson haws, that to English eyes may give 'a Kentish winter aspect to the valley. Perhaps it is the character of these respective pursuits that has given to each of the Gipps Land towns a distinctive individuality—Bairnsdale, young, lusty, and energetic; Sale sitting quietly prosperous on the margin of its green plain, but wanting perhaps an Atlas to hoist the place upon his shoulders and keep it in the front rank of progress. One local edict, which must please a lover of nature, forbids the use of a fowling-piece within the precincts of the borough. As a consequence, water-fowl are thick about the weedy pools, and hundreds of red-legged coots patter along the margins of the lagoons within a few paces of the cottage doors.

Gipps Land might be called new Scotland, for names that are sacred to the Gael are met in every corner of this "land of the mountain and the flood." The adventurous Highlanders who pushed down here from the Manaro country exercised their pioneer rights to the full, although many parts of the district now bear titles given at a subsequent and more formal christening. And in this land of many streams there is no river trip more pleasant than to steam up the Tambo in the early summer. The palisade of yellow reeds that line the banks, drawn inward by the movement of the steamer, bends forward in a stately graceful courtesy, and is then tossed rudely back in rustling resentment by the wash of the sturdy little boat. Farther up the white cliffs rise high on either side, flecked with patches of yellow wattle bloom, and stringy barks wave their long pendent branches from niches in the white walls. There is, perhaps, one stream in Australia more beautiful. The Gellibrand, whether leaping down from its hundred fountain homes away up in the solitudes of the Otway range, or flowing under its green arches to the sea, has no rival amongst Australian rivers, but at last it rolls sluggishly through dark marshes as uninteresting as the clumps of ti-tree that cluster about the mouth of the Tambo. The Gipps Land rivers have peculiar interest for the geologist, and for any one who will note how natural forces have been here excavating and there building up, century after century. The mouth of the stream would naturally be sought for at some spot in the lake's rim where a break in the distant range seemed to indicate the point of convergence for a water-shed. But with the Gipps Land rivers, year after year, the flood waters brought down the surface soil from the ranges, and as this was cast on either side the river gradually crept out into the lake, still rolling between its low banks of mud. And as these banks were raised by the never-ending succession of floods, the lower vegetation first established itself, and then the eternal gums. Thus two narrow lines of green pushed out into the middle of this inland sea, with the river still confined between them. The Mitchell, which first entered at Eagle Point, has been thus embanked out into the lake for miles. Steaming up the stream, one may toss a cinder over the tree-tops on this strange jetty into the lake on either side. When the first boating party skirting the lakes in search of a river mouth found themselves cut off by a band of green turf jutting out for miles, their astonishment must have been great indeed. This invasion of lake territory has not been rapid. An old inland navigator may remember when that dead tree-trunk, now almost buried in silt, was brought down by the yellow waters and cast up high and dry at the river's mouth. The intrusive ti-tree has crept down very near the point, and the gums are treading closely upon its heels. Those for whom anti-Mosaic calculations have a fascination may roughly estimate the thousands of years during which the Mitchell River has been building up the green walls that stretch downward from Eagle Point.

In crossing Lake Wellington a fine amphitheatre opens to the northward. The dark line of ti-tree in the foreground yields to a pleasant green, which in turn fades away to hazy blue when the prospect closes in the long semi-circular sweep of mountain. Every cone and peak in the blue crescent wall has its own name, but the symmetry of these Gipps Land mountains is, if anything, too faultless. Neither earthquake nor volcano has played its ponderous pranks amongst the hills. To identify the spirit of upheaval in his Australian haunts one must visit the Grampians of the Western District, where the ranges, all rising gradually on one side, seem to have been abruptly broken off, so sharp is the opposite slope.

Gipps Land has become such a popular "summer outing" place that its features are familiar to Australian tourists. The narrow strip of country running between the ranges and the sea, from the Lakes' Entrance down towards the Snowy River, is, perhaps, least known. It was a favourite haunt of the aborigines in pre-settlement days. Now a black patriarch, coming back to his old home under the range, might say—

> "A change we have found there, and many a change,
> Faces and footsteps and all things strange,"

though to the eye, weary of bricks and mortar, it yet retains its savage charm. Of a winter's morning amongst these southern woods, the frosty air brings out a rare perfume from tree and shrub which we never find on those scentless summer mornings when the sky above the range is yellow with the flush of the sun-streamers. The tender shoots of one of the shrubs is sticky with a sweet-smelling gum, the rich aroma of which fills the glades. Surely the pharmacy of the future will include many drugs and perfumes distilled from these fragrant bushes.

In the valley there are many indications of native camping grounds. Indeed, the skeletons of huts that have been recently used yet remain—fragile illustrations of primitive house-building. In one of them, pyramid-shaped, the cross-piece rests on forked saplings, while in a circular hut, the uprights, inclining to a common centre, complete the figure of a cone. Only the ribs are left, for a fire-stick—the great native scavenger and disinfectant—was thrust amongst the thatch when the huts were abandoned. Indeed, there are traces of fire all along the slope that suggest a line from " Bush Ballads "—

"All fire-flushed when forest trees redden on slopes of the range."

In amongst the ti-tree there are occasional pools holding a few gallons of sweet, clear water; and some of them almost hidden in marsh-weeds. These were, no doubt, scooped out by the natives long ago, but to-day they serve as drinking fountains for the wallaby and wombat only. There is no mistaking the beaten track leading to the pool for a sheep or cattle track. Just above the trail the space is clear, but at a certain height the bindweeds and tangle bend over and entwine, so that instead of an avenue there is a rustic colonnade. Up in the north the natives had to store water in the hollow of tree trunks, and sweeten it with the blossoms of the banksia and wattle, but nothing of the kind was needed here. Upon the trunks of fallen ti-tree there are masses of scarlet fungi looking very bright in the gloom, and about the bases of the growing trees long brown mosses are creeping. Tree trunks, that have the outward appearance of being sound, crumble away under foot. The springy soil itself is the dust and ashes of dead tree giants. These light straight saplings furnished spear-shafts for the blacks. Here along the coast they were accustomed to adorn the "spear-thrower" in barbaric fashion with pieces of bright shell glued on with acacia gum. The patches of dead brown on the boles of the gum trees show where the natives cut their bark canoes. The form of the canoe is indicated, and a convenient bend in the tree to suit the shape of the canoe was generally selected. Where settlers remove the bark for roofing it is cut in formal squares from straight trees.

Tribes moving up the lakes started from this point, and one can picture to oneself how many thousands of bark canoes have been propelled against the swift current of Reeves's River up into the broader waters. The blacks, instead of crossing, always coasted round the lakes. These fine old conservatives changed never. Their bark canoe, designed by some black Abraham, was handed down without alteration, and its build will die out with the race. They were landsmen pure and simple, and by comparison the Maori, the Fijian, and the Papuan are very Vikings. If sea-kings were few, there were many forest kings about the lakes, although the monarchies have all commingled into the fragment we now see.

A look-out from the summit of one of yonder ranges commands a long stretch of the Southern Ocean. It may have been from this point that the great white seabird, towering as high as the gums, was first seen ploughing through the waves. On the summit of the knoll there is a cairn built for the purposes of a geodetic survey. Sitting under this beacon in the moonlight, one may, on this magnificent winter night, people the glade between the white line of surf and the band of shadow at the foot of the slope with the generations that are buried down among the trees; but in summer the strains of modern music and the beat of dancing feet are carried faintly up from the point with every whiff of sea-breeze. It was little wonder that the blacks loved the quiet valley, and made it their summer home. In addition to the spoils of sea and lake, hundreds of kangaroo in the dry seasons flocked down from the range. The native hunters had that variety of fish, flesh, and fowl that the city sportsman with his patent choke-bore and jointed fishing-rod finds to-day. The trees from which the natives got the white grubs —to them as great a delicacy as our oysters—have all been killed in the interests of trade. But the blacks of this country had once their oysters too, for beds of large shells, beside which the Sydney rock oysters and Stewart Islanders are puny dwarfs, have been found on the banks of one of the Gipps Land rivers. The wild dogs of Gipps Land are as fine specimens as could be wished for—big, prosperous fellows, with the square, stolid-looking heads of their kind. The dingoes, even in captivity, never lose their wild native look. Their sharp, black, restless eyes are not those of an animal easily tamed. Even their one-time masters, the aborigines, soon discarded them in favour of the more affectionate domestic dog. And although when taken young they live in peace, if not altogether in friendship, with the blacks, no amount of taming will reconcile them to a white man. Perhaps the finest specimens of the native dog in Australia are found in the ranges running parallel with our southern coast line. Those who have camped in amongst these hills at night have heard their wild howl waking up a hundred melancholy echoes in the deep-wooded valleys.

Down the coast there are tall cliffs, about which the sea shrubs creep like juniper bushes, not daring to raise their heads in the face of that ninety miles of storm and surf. In the damp niches of the cliff there are patches of young moss that shine like emeralds set in a shield of bronze, only the moss is brighter than the brightest jewel ever shaped by lapidary's wheel. In the underwood black wallabies are darting about, and their footprints are plentiful about the sand hummocks. Here and there a larger track, with the great central toe sunk deep in the sand, shows that a kangaroo has been down from the timber in the night. The headlands are interesting to a naturalist. The bank is thickly studded with coloured pebbles, representing many varieties of rock, and these have been washed out and rounded by the tides until, lying here on the beach, they form the material for as handsome a garden walk in rough mosaic as could well be imagined. Similar pebbles, some of them very beautiful, are found on the coast near Moonlight Head.

The lakes have an additional charm in the fact that much of the wild *fauna* yet survives. At every stride in the valley parrots go screaming away amongst the trees, and as they cross the sunny spaces there is a flash of dazzling green, with suggestions of orange and crimson. Under the ferns at the foot of the slope the bank is honeycombed with wombat holes, but most of the dwellings have long been vacant. The wombat is very powerful,

F

and can turn a boulder almost as large as itself out of the way when it bars the road. At some of the state plantations on the ranges large stones are often placed in a gap in the fence, but these were generally found thrown on one side by the wombats in the morning. Their natural inability to defend themselves either by flight or fight has compelled a nocturnal life, so they are more often heard than seen. At night, when fishing in some of our mountain streams, I have heard them fossicking and burrowing in the scrub about, but have never caught sight of one. Wombats are plentiful amongst the Otway Mountains and along the Dividing Range. On the farther side of Mount Disappointment there is a spur of the range literally honeycombed with their long winding tunnels.

From one of the wooded headlands a sea eagle has sighted a flock of black coots out on the lake. As he "quarters" above them the coots rush here and there, beating their wings upon the water; the sound comes as though a hundred swans were taking wing. The eagle winds down and down, and the coots scream and huddle helplessly together. But scream, and dive, and cluster as they may, one of them must die. The master bird is within a few feet of them, and still he sails leisurely about, prolonging the agony of the victim beneath. Then he strikes—never swooping like the black hawk, but lazily—and flies away to the headland with a screaming, struggling coot in his claws. To-morrow the function will be repeated, but the constant slaughter has no apparent effect in thinning the ranks of the water birds. In a single flock, stretching for a mile across the lake, there are countless thousands of coots. Sometimes the lake harrier carries away a sea salmon or a fat perch, and in their capture his wing power is shown to better effect. You find the skeletons of the fish lying on the high points of the different promontories. Up in the rivers the waterhen, or porphyry coot, with bright red legs and bill, paddles about among the rushes or flaps lazily upward nto one of the bushes that fringe the stream. Passing through the straits that connect the lakes, broad stagnant pools speckled with yellow marsh flowers open out, and here hundreds of coots and swans, with occasional blue and white cranes, make up a highly interesting bird community. Far out on the water the big pelican sails in solemn single state. They were safer years ago, for the blacks neither destroyed the bird nor its eggs. Not through any veneration, however, but simply because a pelican diet was too rank even for the aboriginal palate. These big birds take wing in a singular way, not patting the water one foot after another, like the swan, but by a succession of leaps, striking the water with both feet at the same instant. The heights above the entrance are a naturalist's paradise. From the denser thickets wonga wonga and bronze-wing pigeons—shy as a nightingale—flutter out, and are gone in an instant. The native thrushes have a plumage as bright as the bronze-wing, and with their white throats and fantail —a characteristic feature in tropical birds—are altogether more gay than the modest, sweet-voiced immigrant thrush hiding beneath the laurustinas in our city gardens. Down in the grassy hollows one is startled with the sudden hum of beating wings, and a little king quail, not much larger than a pipit, whirrs away into the distance. Some of the little islands out in the lakes are a nursery for quail. The birds let loose there never take the long flight to the mainland, and multiply amazingly. These islands are as strictly guarded as a Highland deer forest or an English game preserve. Amongst the most beautiful of Gipps Land birds is the white crane, with its curved neck, small serpent head, and snowy plumes flowing out airily from back, breast, and head. The bittern, with its wealth of brown breast feathers, drums in the starlight, and about the lake timber the sooty owl, in appearance an undoubted nightjar, but with talons almost as terrible as an eagle, makes his home. When the rustle of his softly-beating wings was heard in the darkness above the aboriginal camps the natives whispered to each other that a bad spirit was abroad. Seabirds are thick about the shore. Pacific gulls —dull grey, of black and white—wheel about in hundreds, and when a dead fish is cast up by the waves these sea scavengers, with their strange razor-bills, soon dispose of it.

On the Gipps Land Lakes feeding the sea-gulls is a popular amusement. The birds wait for the Bairnsdale boat at the entrance to one of the passages connecting the lakes. Long before the steamer is in the straits the flock of silver gulls, with perhaps a pelican amongst them, can be seen waiting on one of the sandy points. It is the hour for luncheon, and the gulls know it. Almost before reaching the water tasty morsels are caught up by some of the long train of birds flitting in the wake of the steamer. The passengers spend a few pence in bread, so that the innocent pleasure may be prolonged. To-day a bishop is casting his bread upon the waters. The birds, becoming more confident in the competition, venture within a few feet of the hands feeding them. Sometimes the bread is cleverly caught in mid-air, and if there are cricketers present the bird is applauded. Each as it secures a prize falls out of the rank, and, flying away to one side, finishes its meal alone. Otherwise the pleasurable proceeding would be shortened to a scramble, for in this democracy possession carries with it no state guarantee. The birds accompany the steamer right through the lakes, and meeting the twin boat on the down trip, sagaciously turn and become her escort.

In steaming down the Gipps Land rivers, funeral cockatoos pass constantly overhead—the spots of flame in their broad tails lighting up their black plumage. The woods are noisy on either hand with the shrill call of cockatoos and parrots. Some naturalists separate the family, thus, one might say, giving patient, dignified Nature a rebuff by inventing divisional lines for her. The bushman who studies Nature in its own solitudes classes them all as cousins, and names the connecting link a cockatoo parrot. There are four commonly known varieties here: the big white cockatoo, with its handsome saffron plume; the corella, or as some country people libel it "gorilla;" the rosy cockatoo, resplendent in deep pink and ashen grey; and the creamy-white cockatoo, with a faint flush of crimson showing through its plumage like the Aurora Australis in a clear sky. After the late summer rains, when the plains of Riverina are covered with broad, shallow swamps, the cockatoos congregate in the rushes about the margin of the water, and then the bird-catchers with their decoys and white-tilted waggons push out from the border towns to thin down the bird clouds. White cockatoos are nowhere seen to such advantage as amongst the red gums, where a large flock with its beating wings convey the idea of a living snowstorm. Although big cockatoos are such sociable fellows when raiding the wheat-stacks in company at daybreak, they fight desperately when for the first time imprisoned in the bird-catcher's cage—each bird apparently putting the blame for their common misfortune upon its fellows. Sometimes the bird-catchers are themselves trapped. One of them camped in a sheltered hollow on a stormy night found the water pouring into his waggon, and was driven first to the roof and then to the branches of a myall by the rapidly rising flood. The Wakool had burst its banks in the night, and the yellow waters had stolen quietly out in the water-courses—as you may have seen the fog creep down from the hills in the early morning—and invested the plain. The bird-trapper remained for twenty-four hours in the tree-tops. The parrots range from the big King Lory—a corruption, of course, of the Spanish word *loro*—that wakes up the echoes in our mountain-tops to the little yellow and green shell-parrot, delving like a honey-bee down amongst the petals of the waxen gum-blossoms for the nectar distilled there. The screech of the rosella, even if heard in the smoky city, suggests box boughs and musk trees. Those who have seen the rich primal colours of the Lory thrown with relief against the dead gold of wattle blossoms never wish to see the same bird in a cage.

The Green Leek should never be kept in a cage; it has the tongue of a honey-sucking bird, and is quite unfitted by Nature to live on a diet of cracked maize and flattery. The eagle hawk is found in Gipps Land as in every other part of Victoria. The qualifying name is misleading, for the bird is as daring and rapacious as most of the recognised eagles. I know one instance of a pair having killed fourteen full-grown turkeys from a flock out in the western district. They once lived largely on opossums, but the pretty little ring-tails have been much thinned out by these fierce hunters. Even in its neat round nest the opossum is by no means safe, for the eagle with its strong claws tears away the covering and flies off with its struggling victim. The curlews or stone plovers, inoffensive as they look when daintily stepping about some retired glade where for years their home has been established, destroy many young game birds. In the season you rarely find the nest of a quail or a pipit in the same piece of ground that the curlews have "pegged out." They build year after year in the same patch of grass, and send their young abroad into the world to seek fresh fields. They are very vigilant, and are seldom sighted until they have first taken alarm, and are running rapidly through the long grass with head and neck bent low to escape notice. If not too closely followed so as to force them into flight they are easily "driven," and thus the birds are shot down. The eggs are seldom found of exactly the same shade, and the colour is generally in strict harmony with the grass tufts about the nest. You may step within a few inches of the nest and fail, except by the merest chance, to notice it.

A flock of a hundred native companions in one of their grotesque quadrilles is a strange sight, but the moonlight minuet of the lyre bird is more interesting still, and far more rarely seen. Along the Gipps Land hills that stretch from the Tambo westward to Lake Tyers, and overlook the sea, the lyre bird has his haunts. In other parts of Victoria you find this interesting bird, but in few places more plentifully than amongst these lonely peaks. Up in the Dividing Range, where the mountains are tapped by the white stone aqueducts that bring water to Melbourne, the birds are seen rushing across the narrow bridle track, but very rarely. The lyre bird is at his best when seen at home, and, like many other reserved people, only exhibits his most agreeable social qualities there. It was on the occasion of an outbreak of one of those periodical gold fevers that so agitate Gipps Land, and when the discovery of a bit of water-worn gold in some wooded gully is accepted as an unerring fingerpost pointing the way to other "Long Tunnels" up in the range, that I had the privilege of an introduction to the lyre bird.

A streamlet, as it trickled century after century to the sea, had cut a deep chasm in the range. From both banks hills rose in sharp slopes covered to the summit with scrub, but on the western side there was a little open glade, which had been pointed out as a night playground of the lyre birds. Here, in anticipation of such a gathering, we sat under the shade of the trees on the eastern slope waiting for the moon. Away out on the ocean there was a broad band of shifting moonlight, but from our recess the first rays were alone visible through the dark trelliswork of trees on the crown of the range. Then the western slope was gradually illumined, and the light crept down until we uninvited guests in the shadow could almost count the grass blades that carpeted the lyre birds' hall, and could see the beautiful liquid glow of dewdrops on the turf. In the bush noises that quickened with the growing moonlight my fellow-watcher—a practised bushman—fancied he detected the mimicry of the lyre bird. Its own cry is a peculiar liquid gurgle that suggested the native name of "Bullan Bullan," but the bird is an admirable mimic, and perhaps the strange whistling crack of the whip bird, or the melodious metallic tinkling of the bell-birds, are the only sounds he cannot imitate. On these bright nights the croaking of a marsh frog, the grunt of a wombat, or the screech of an opossum, may all come from the versatile throat of the lyre bird. In watching for rabbits at night you rarely see them hop out of the burrow, but find on the dusty mound in front of the home a little sentinel sitting rigidly on his haunches, as though he had relieved guard two hours before. So with the lyre birds. Our eyes had been cast for a moment up to the western mountain-top to note the effect of that brilliant moonlight amongst the leaves, and when we looked next towards the little glade it was no longer vacant. Out in the open was a beautiful male bird with drooping wings and tail thrown forward so that the head was entirely hidden. He stood for a little while stooping slightly forward as a man might when listening intently for the coming of a friend or a foe. In the light the successive black and white bars of the classical tail were distinct, and in fine contrast with the dull grey body. The bird was not long inactive. First he drummed on his mound, and then he waltzed round in a sort of ambling canter, as though controlled by some magnet in that little hillock. He seemed to have practised every pirouette of a modern ballroom, and had some ballet dancer of repute watched this vivacious bird at play she might have found some new notions for an opera bouffe or a pantomime. To describe his shifting movements in detail would be impossible. Some were graceful, some grotesque, but every action was that of a bird keenly conscious of the delights of life and motion. A second bird soon joined—a pair of right joyous fellows dancing there in the moonlight.

Sometimes the lyre birds are shot while thus absorbed in their dance, but the man who would fire upon a little family party of this kind deserves, as a mere matter of poetical justice, to have an accident with the second barrel. At the close of last winter a couple of young bushmen, while out shooting lyre birds, were lost for days in the ranges. It is to be hoped that they reflected on their enormities.

We have our bush lizards too. The iguana, startled from its haunt in the grass, rushes for the nearest tree and spins up it merrily. Once on the tree-trunk it is impossible to shoot the iguana while in motion, for however rapidly you may move round it the bole, somehow, always intervenes. Even the youngest iguana is found to descend the tree on the opposite side to the intruder who has startled it. Curiosity is its failing, and sometimes, at the very threshold of its home, it stops to have one more look at its enemy. When the iguana is dissected, two large masses of yellow fat like butter are found, and this, reduced to lard, is a popular healing ointment with some bushmen. They are great destroyers of vermin, and a comparatively slight nuisance in other respects, so that the authorities would act wisely by giving them a perpetual protection with the lyre bird. The blue-tongued lizard is not by any means rare, and may be found in scores wherever there is a bit of rocky country. When out sunning themselves they have an awkward habit of lying quietly until you get between them and their retreat, and then making a sudden dash for home almost between your feet. With any one who has had a few snake scares this always induces a momentary gymnastic display. The coast lizard is a dark-coloured reptile, almost as large as an iguana, and ten times as ugly. Some of its progenitors seem to have suffered an internal dislocation, and the catastrophe has been faithfully handed down. It presents the strangest collection of acute angles ever put together in one animal. There is no harmony in its outline, every joint clashing with its neighbour. It is quite harmless, and if only a bit more handsome would be as well worth preservation as the iguana.

One cannot look upon this Gipps Land country without feeling how different are the conditions of life here and on the sunnier plains of the Wimmera in the north, where the toil of thousands is that heart-aching toil that knows no reward but debt and disappointment. When will these northern farmers grow weary of so much thankless tillage, so little harvest? Will the homesteads that dot the plains be one day deserted like the eagle hawks' nests one sees in the dead gum trees. To further the comparison, it may be said that the vegetation of these two corners of the colony is typical of the measure of success that in each has attended man's enterprise. The stunted oak groves of the Wimmera, in appearance something between southern sheoaks and northern pines, are sadly suggestive of universal poverty; while the stability, the variety, and the grandeur of Gipps Land forests typify her many and lasting resources. May the anthem of peace and plenty that the Gipps Land settlers sing to-day last always!

Juvenile Poachers.

POACHING in Australia is but a mild imitation of what in England was once a heinous crime. In the old world the game laws have been productive of entertainment for some, but bitter misery for the great majority. And Australia perhaps has profited more than any other country by the rigorous provisions that amongst the hazel thickets of England, the waving heather of the Scottish mountains, or upon breezy Irish hills, made "wealth accumulate and men decay." In the philosophy of the new world no social stigma rested upon the man whose only sin was the snaring of a hare or pheasant at night, in some grand demesne, as a resource against the poor-house. After all the old-fashioned poacher was very much the sort of man required to fight the hard pioneer battles of a new country, and many successful Australians of to-day, who under the Southern Cross have worked their way to honourable wealth and position, were looked upon and treated as criminals in England thirty years ago. Sympathy with the love of forbidden fruit in the way of preserved game is probably the out-cropping, in very pronounced form, of an inherited love for hunting and field sports, which (in my own case, I suspect) had its origin ever so many generations ago in the chasing of red stag and roe up amongst the crags of Glengarry or about the rugged shores of Loch Nevis, far away in the Highlands of Scotland.

When hares and rabbits, with a few lesser evils in the same line, were turned loose upon civilisation here, they spread very rapidly along the valley of the Deep Creek. Many of the landholders, pleased with such reminiscences of rural life in the old country, preserved game strictly. Had they been familiar with rodential natural history, the effects of such fostering care—in a land where there were no heavy snowfalls nor any Malthusian checks to rabbit population—to check the poachers would have been encouraged in Gipps Land. As it was, however, both poacher and game survived, and in after years, when the latter became a much worse nuisance than the former had ever been, invitations were issued freely to shoot all and sundry.

On slopes where on moonlight nights I have nestled in the grass waiting for the solitary hares that came cantering down from the white plains above to feed on the sweet young grass in the river bottoms, everything is changed. Crack shots from the metropolis with improved breech-loading choke-bores, in company with those unfortunate casualty sportsmen, armed with relics of the Queen Anne period, now wander in broad daylight, and shoot as they please.

There was a charm in those night expeditions, outside the excitement of the sport or the risk attached to the trespass. A charm in the beautiful Australian night, then almost unconsciously felt, but realised more fully in later years. It was not strictly conducive to success in night-shooting to sit in the shadow of the stone wall with a bright gun-barrel glistening in the moonlight in a way to make every hare in the district give it a wide circuit, while one worked up shadowy fancies about the stars, and watched for the new twinkling points that came out in multitudes between the brighter orbs as the night deepened. Neither was it "business" to give oneself impromptu problems as to how often the stone plover would call in his next shrill piping from the wattle scrub, or to guess the number of cormorants in the next wedge-shaped flock that came sailing across through the gloaming on their way to northern waters.

Yet all these were pleasant features in the sport. My armament was more antique than effective—a long-barrelled shot gun, the present of an old farmer, who had himself learned to use it somewhere on the Weald of Kent. The original trigger had disappeared, but a roughly-forged horseshoe nail, that jagged one's finger terribly, had been substituted, and in some way had increased the pull fivefold. It was a moot point amongst a select circle of juvenile "sports" whether the killing powers of this weapon were equal to those of "Old Anthony"—a sort of modified long-range cannon held in general veneration throughout the district. The weapon in question was an Enfield rifle, which its former owner boasted had done service in India for the Queen, by shooting rebellious sepoys. On account of the rifled barrel, it never by any chance threw a charge of shot twice in the same direction, but was frightfully destructive when emptied amongst a flight of golden plover or wood-duck.

There was also a family relic in the shape of a six-chambered revolver of obsolete pattern—a sort of reminiscence of the firs and hemlocks of Upper Canada, from whence it came. This was kept largely for show, on account of a difficulty in inducing it to explode with ordinary gun caps. At rare intervals, and during casual spasms of industry, these weapons were burnished until they shone like a newly-minted florin, but after a season they always relapsed into a chronic condition of rustiness. Indeed, it became a tenet that the rusty barrel was best, for the chalk mark which served as a substitute for the sight was more easily discernible thereon either in the darkness or in the moonlight.

Juvenile poachers generally worked in pairs, for the sake of companionship. Sometimes it was terribly lonesome on the plains, and one was apt to get pondering on all sorts of possibilities, and, as a consequence, go home sooner than necessary. There was a world of pleasure in the recital of our prowess to the schoolmaster on the following day, for he was as keen a sportsman as ever grasped a gun-stock or a bamboo fishing rod. In these cases, however, prudence always suggested a slight change of locality, in order to give the excursion an air of legality. Let me recall one of these night trips.

We stroll out of the village in any direction except the one we wish to reach, and once under the shadow of the encircling hills wheel away to the right and towards the upper end of the valley, where the young crops are just springing from the red soil. We must keep below the level of the sky-line, for against the red glow of the sunset in the background our figures would be clearly outlined and visible to every soul in the village below. At the stackyard near the foot of the hill a brown-backed, yellow-legged kestrel is industriously beating the bushes for sparrows and mice, and creating quite a sensation amongst these small game. In the paddocks of English grass on the opposite bank of the river a long line of well-conditioned shorthorns, of the Booth strain, with most ancient pedigrees, " winds slowly o'er the lea." There is a heavy dew upon the ground to-night which will have hardened into a frost before morning, but the damp grass is only a minor inconvenience. Now we are between the young crops fringing the river side and the uplands, the two connected by a gently-falling slope, at the bottom of which the " station " buildings lie.

At the very bottom of the slope there is a still more abrupt fall, and on the summit of this the hare must be shot, hence we select our post at a range of from twenty to forty yards. Then we sit down and wait patiently for the game. After a few minutes the spot that had seemed entirely deserted when we first came shows signs of life. A spotted native cat, out poaching on his own account, after a quail or a ground-lark, stops suddenly within a few feet of us, and begins to study, in a half-comical, puzzled way, this new feature in the landscape. The suspicion of a movement, however, makes him scamper away with a rattle amongst the closed Cape-weed flowers. A magpie perching in a sheoak close by wakes up for an instant, and mistaking the twilight for the dawn, breaks into a hesitating carol. Then he realises the position, and tucking his head in humiliation under his wing, goes off to sleep again. Now a dark object comes into the field of vision, moving swiftly towards us down the slope, and halting at intervals to reconnoitre.

This is the game for which we are watching. Those long sensitive ears catch the very slightest noise, and every sound is invested with suspicion. The hare reaches the little horizon that we have ourselves arranged, and for an instant its form is distinctly outlined against the glow that lingers in the west.

"Hist!" The hare has caught the sound, slight as it was, and stands upon its hind legs, rigid and erect against the sky-line. Aiming for the upper edge of the broad patch of white, you fire. A half-human cry wakes up the echoes in the gullies, and one feels terribly guilty just for an instant, as the poor victim kicks out its life amongst the folded buttercups and pink-streaked field daisies.

There was an element of moderation in our sport, and after shooting one hare we seldom waited for a second. Remembering the greyhounds at home, we perhaps potted a couple of the opossums which all night long were holding noisy festival amongst the gum trees down by the river side, or perhaps strolled up to the horse paddock to see if the curlews were on the move, and tried for a chance shot. Often, when the night was young, we had an hour at the white sandbank, where enterprising colonies of rabbits had formed their burrows, until the face of the limestone cliff looked like a great piece of empty honeycomb.

It was strange that, watch as closely as we would, we seldom saw the rabbits in the act of popping out, At one instant the cliff was all an uninteresting white blank, and in the next a big brown rabbit was sitting in front of one of the burrows as rigidly as though he had been on sentry there all night. In yet another instant he was turning somersaults down the bank, and the report of the gun was being repeated for miles up the gorge.

The zeal of the landholders for the preservation of game vexed anglers also. Gun and rod were jointly forbidden, and thus long stretches of water, some of it favourite fishing ground, had to be worked in secret. Discovery here merely meant, however, a request to prospect in other waters, coupled with a warning about coming back again.

After the spring floods black fish and eels were busy all through the day, and then those in whom angling inclinations were strongly developed found a double delight in dropping a line into shady corners of forbidden pools, and bringing up big sluggish blackfish that had prospered under protection.

The size of these fish varies with the locality. Where the stained trunks of trees that have been submerged for perhaps a century were thickest, there were the black fish in the midst of them. Away in the crevices between the logs, where a line was cast only with a deft hand, and by the most delicate manipulation alone kept clear of innumerable snags, families of black fish had lodged for ever so many successive seasons.

Down amongst the aquatic plants at the bottom of the deeper pools silver eels glided in and out amongst the weeds, " fossicking " for choice morsels in the rubbish brought down by the stream. In these surroundings there was everything to please the senses, and an amateur naturalist could wish for no finer field for study.

Black wattles drooped down from the high banks on either side. Half-way up the crags, on the opposite bank, a great mass of sarsaparilla plant, its blue blossoms looking quite radiant in the sunlight, clung to the rocks, a tempting sight for a botanist. This flower, however, fades even more rapidly than the majority of wild flowers, and ten minutes after a spray is plucked its glory is gone.

There are two species of the sarsaparilla here—the one medicinal, the other merely beautiful. It has been said that the former is found only on the north of the Murray, but here it is flourishing considerably to the south of the Dividing Range. Nature seems to be holding her spring flower show. The wild convolvuli, pink and blue, wind about the reeds, and down at the water's edge there is a perfect carpet of little white and yellow marsh-buttons. In the centre of the pool a couple of pink and white water-lilies, with their delicate waxen petals showing to full advantage in the light, help to break the monotony of yellow, the prevailing colour among the wild flowers here about.

Under the shadow of the broad green leaves, the trout that at our approach shot like a gleam of silver from the sandy shallows, has taken shelter. You can just fix his position by the gentle sweep of the broad tail, or the oscillation of the fins as he waits patiently for the insects that may come tumbling down with the current, and float towards him.

A yellow grasshopper is drifting about in the back water at the upper end of the pool, sometimes partly caught by the current, and then wheeled back again under the banks. Now it comes fairly into the stream and floats towards the water-lilies. The trout rises quickly, there is a swirl on the surface, and when the ripple clears away he is back in the old spot under the lilies, the grasshopper inside him.

It is rather unfortunate that from one point of view the "tastiest" will not "survive" here. Nothing is better eating than a properly-cooked blackfish. The English trout are annihilating them, however.

Sometimes, after running all sorts of risks to gain a favourable spot on the creek, fish wouldn't rise a bit. Perhaps there had been a flight of beetles, or grasshoppers, on the previous day, and appetite was stalled for the time being. There was always something, however, to look at. At one time it was a blue heron fishing in the shallow water with more success than had attended our efforts in the same line, or a blue coot with its bright red bill and legs pattering along amongst the sedge and rushes on the opposite brink, and now and again dashing for a dainty morsel stranded in the shallows.

On the top of the ridge that *rara avis*, a black cockatoo, is hard at work drilling the soft bark of a "messmate" tree for insects. If ever a bird earned its living surely this one does. The bark hangs in shreds all the way up the trunk, and sometimes even the hard wood is chipped and splintered. An American, on inspecting the job, would conclude that the woodpecker, who has whittled away whole pine forests in his own state of Maine, had been trying his skill on the tougher timber of an Australian forest, and would be surprised to learn that it was all the work of a pair of sable cockatoos. If one of these industrious artisans was placed in a wooden cage at night he would have eaten his way to liberty before morning. If he had no opportunity to thus use that destructive beak it would in a very little time grow so rapidly as to give him lockjaw. The rain-bird is calling in the thickets today, a very sure sign that the beautiful spring weather will not last much longer. An interesting little brown fellow, with a modest preference for hopping about in a retiring way amongst the denser bushes, his note is a better indication of the coming rain than the barometer. How the little weather-prophet gleans his information on coming events is past comprehension, for to the weather-wise there is no sign of storm or tempest in the quiet serenity of this spring day. Then attention is turned to a pretty bronze-wing pigeon which has fluttered out from the scrub on to the bark of a giant gum tree, and appears to be admiring the prospect. Similarity in form, but distinction in colour, marks the different varieties of the pigeon tribe. Few birds are clothed in more conspicuous or varied lines than the several domestic varieties, and in the country none fall easier victims to their destructive enemy—the fast-flying black hawk. On the other hand, no birds have their markings so evenly distributed, or show less variation in plumage, than the several varieties of our beautiful southern wood-pigeons. The habits of both wild and tame birds are somewhat similar, yet while the former, with its bronze coat assimilating so closely with the hue of nearly all our forest trees and shrubs, continues to prosper, its civilised compeer could not, under the same conditions, escape annihilation. A pigeon-fancier would in time develop some remarkable colours with the pretty bronze-wing sunning himself on the tree-trunk yonder, but if the "improved" bird regained its liberty it would be just as attractive to the eye of that other pigeon-fancier—the black hawk—and with results much less satisfactory to the pigeon.

There, on the trailing wild raspberry, is "a character" amongst Australian birds—a cuckoo in miniature and manners. No domestic anxieties disturb its peace of mind during the nesting season, for after the fashion of its British prototype, it has left one of its chocolate eggs in each of the nests of its neighbours the wrens, who twitter and flit amongst the bushes as though they esteemed it a high honour to provide for the wants of this alien. Are wrens colour-blind and unable to discriminate between the dark-chocolate egg of the cuckoo wren and their own snow-white shells, or is the foster-parentage a matter of mutual agreement founded on treaty? Once, as a boyish freak of mischief rather than an experiment, I changed the eggs of a black-and-white wagtail and a brown groundlark, but the parents accepted the position with the utmost good faith, and reared each other's families cheerfully, though they must have often been sadly troubled in striving to discover that likeness between themselves and their offspring which, it is said, all parents are so anxious to trace. The substitution of one egg for another in the nests of a magpie and crow was also received with perfect confidence, probably as a freak of Nature.

Only one experiment of this kind turned out a disastrous failure. Hunting one day in the thistles we flushed a fine mountain duck from her nest, and found about a dozen eggs all snugly placed in that warm bed of down. We carried away half the clutch, and gave them to a hen, who, after a season's secret laying in a bunch of wild chamomile down near the pond, was then engaged in the barren task of trying to hatch chickens from an india-rubber ball and the bowl of a ruined egg-cup. After due time had been allowed for incubation we looked at the nest again. The egg-shells had just been pushed aside when the little mountain ducks burst their way out, but the interesting strangers themselves had vanished for ever.

Having given a few incidents in the career of a juvenile poacher, a short sketch of his two chief companions may not be out of place. Few dogs have shown more complete sympathy with the tastes of their master. Like him, they knew every corner of the valley and every wooded bend of the creek for miles, and they were more familiar with the rabbit-burrows of the district than with their own kennel.

In addition, they were as hopelessly vagabond and thoroughly useless a pair of curs as ever evaded the dog tax. The sight of a gun was a greater pleasure to them than the choicest canine dainty, and at a very early age they learned to arrange private hunting parties. It was no surprise to meet the pair miles away from home in some lonely gully, where the rabbits had prospered in solitude, and the hares had grown fat and lazy, through too much food and too little exercise. What greatly endeared them to us was the fact that with all their love of sport they declined to work for a stranger. It was in vain that kindred sportsmen, knowing them by repute, borrowed the dogs for a day's shooting; when turned loose, even in the heart of the rabbit country, they either went straight home or went hunting on their own account. I often think longingly of the old days and the old dogs, and these lines, from the American old maid's address to her mamma—

> "Backward, roll backward,
> O Time, in your flight;
> Make me a child again
> Just for to-night,"

have occurred to me, and I have been illogical and sentimental enough to wish that it were possible to see again those vagabond dogs, whose bones have for years bleached white in the gullies where they died.

The best of them, "Tichborne," was a sort of nondescript red terrier of various breeds, who seemed to have inherited all the merits of his numerous ancestors. He was named after the Claimant because it was impossible to trace him to any particular family.

His conceit was out of all proportion to his stature, but seldom has more vitality and resolute canine pluck been put together in so small a parcel. Yet he was withal cautious. A long apprenticeship to vagabondage had taught him that in dealing with some of his enemies courage ought to be associated with discretion. With snakes, his mode of action—or inaction—was peculiar, and led to our riddling many a big barred "tiger" or venomous-looking brown snake that would otherwise have escaped our notice and the charge of duck-shot which we always parted with so cheerfully. Tichborne was never known to tackle a snake, even when it was writhing from the effects of a charge of shot—a time when most dogs lose their caution, and sooner or later their lives. If the snake was coiled upon the grass in the open he moved gently around it in regular circles just beyond striking distance; or if partly concealed by a bush or tuft of bracken he stood in position watching it intently as a pointer would a quail. If the snake was travelling, Tichborne moved slowly backward a few feet in advance of it, always keeping the same distance ahead, and with his eyes fixed steadily upon the snake. At first we attributed this conduct to a possible power of fascination, about which a good deal has been written on rather slight foundation, but finally agreed that although Tichborne was too cautious to come to close quarters with what his instinct seemed to tell him was his most dangerous enemy, his nature would not permit him to pass on and leave the snake in peace. Most dogs either rush in as soon as the snake moves, and shake the life out of it—getting bitten in the *mêlée*—or circle round it barking, and emboldened by the apparent apathy of the reptile, get within striking range, only to meet the same fate. I once saw a young dog come upon a big brown snake coiled amongst the rushes near the edge of a pond. After barking for some time, and making repeated feints of attacking, the dog's curiosity seemed to be excited by the fact that the snake, with head raised slightly, and its neck curved in a sinuous arch, was in no way alarmed by this hostile demonstration. The dog moved up a little closer, and then like lightning the snake struck him on the very point of the nose. In twenty minutes doggie was dead. This one had always been reckless in his curiosity, however, and only a few days before his death got too close to an emu which was feeding with her young ones in a quiet clump of timber, only to be kicked some yards away for his intrusion.

Tichborne's caution stood him in good stead—say when he found an opossum almost as large as himself stowed away amongst the gum-tree roots, or in the snug corner of a hollow log, so that it could only be approached in one direction. In defending themselves the opossum and native cat each take a firm hold with their teeth, and keep it as long as life remains in them. When cornered in one of the open spaces of a stone wall, or in a decaying log, they turn their face to the enemy, and with open mouth and formidable mien await an attack. Most dogs at once rush in for what wrestlers would call a "catch-as-catch-can" hold, but Tichborne had been in too many affairs of this kind, and was much too familiar with the punishment received from these sharp teeth,

Bringing his head as close down to the enemy as experience told him was safe, he stood rigid and motionless, with every nerve and muscle strung, every coarse red stubble of hair erect and independent, and a pair of baleful little yellow eyes blazing out a whole broadside of enmity upon the foe. The cat or opossum was always the first to tire of that deadly suspense, and to turn its eyes ever so slightly to one side, on the look-out for some loophole of escape. In that fragment of a second Tichborne went like lightning to quarters, and not even the dart of a tiger-snake could be quicker than his seizure upon the enemy's neck. He seldom missed, and just as rarely lost his hold when once gained, although often obliged to pause for breath. One of his smartest feats, however, was to climb sheoak trees after the native cats which take refuge in the branches when alarmed by night. When the tree leaned in the slightest degree to one side he went up without difficulty, the coarse bark giving a firm foothold. In this way he killed many a wild cat.

"Tichborne" was intelligent enough to become a very useful retriever. He would retrieve rabbits from water, but it was quite beneath his dignity to meddle with a bird. In many places the creek had cut its bed deep into the red loam at the foot of the hills, and on one side there was a high bank, spotted with occasional grass tufts. Here in midsummer, when all the river bottoms were dry and bare, and the fields about were yellow with stubble, the rabbits found tufts of sweet young grass. Sometimes they made their noonday forms in the bunches half-way up the bank, and slept through the sunshine and heat of a warm summer's day. A footfall above them, or a moving reflection in the clear water beneath, was sufficient to send them scampering along one of the narrow pathways to the shelter of a large patch of roses at the end of the cliff, where a whole colony of rabbits had long ago been established. In this race for the rose bushes there was always an opportunity for a fair shot, and many a one we sent toppling head over heels into the water beneath, where, unless their lungs had been pierced by the shot, they floated slowly along with the current. The splash in the water was always sufficient to tell Tichborne that his services were required. He always caught the rabbit by the head, and allowed it to trail over his shoulder—as a fox would a fowl—when bringing it in to shore. Once only I saw him beguiled into swimming for a coot which I had sent scrambling into the water with the greater part of a charge of heavy shot, as the police say, " concealed about his person." Tichborne heard the splash, and knowing that there were rabbits in the locality, swam out to investigate, but on coming to the fluttering coot he turned sharply round again with a very expressive movement, as though mentally remarking, " It's only an uninteresting bird after all."

His disgust at a miss was palpable, and not to be misunderstood; but when himself guilty of negligence it was impossible to imagine a dog more dissatisfied with himself. One Sunday morning we had been for a long early walk, and Tichborne had dallied by the way to interview a native cat in a stone wall. In order to catch up he took a short cut down through the lucerne field, where a sea of bright purple blossoms, loaded with dew, were glistening in the light of the rising sun. Suddenly Tichborne stopped dead, and stood rigid and motionless, looking straight into the centre of a tuft of lucerne, just as you have seen a Gordon setter watch a quail nestling in a burr clump. Then he bounced forward, pinned a hare larger than himself which was lying in its form, and after being ingloriously dragged forty feet down the slope, was left with his mouth full of fur, while the hare, frightened "clean through," went bowling out to the plains faster perhaps than it had ever gone in its life before. I never saw a dog take a mistake so much to heart. As we sat on the fence laughing at his discomfiture he went up at least a half-dozen times to have one more look at that form, and after inspecting it from every point of view, returned each time whining in sheer mortification, and using as bad language as an illiterate dog could be expected to use. Had the hare been squatted head towards him he might have got his pet hold, and would then have been harder to shake off than an octopus.

Tichborne's courage was proved beyond question on one Christmas Eve which I remember very distinctly, because on that day I got very close to the first pair of living foxes I had ever seen. A week before, some one had noticed a very large fox carrying a crippled fore-paw in this ravine, and a couple of days later a dead vixen, which had evidently come to grief over a poisoned bait, was found in the avenue close by. In a rocky gorge at the very end of the defile was a little basin thinly covered with a growth of tall variegated thistle, and here as I was picking my way carefully, for it was genuine Christmas weather, and snakes were plenty hereabout, something red rose from a little bare plot a few yards in front of me and sat upon its haunches, looking inquiringly around. Then a second animal sat up just alongside, the pair sunning themselves after their doze in the heat of noon-day. As I stood perfectly still, anxious not to alarm them further until I had quite gratified my curiosity, Tichborne trotted down the bank and jumped on to a large rock within a few feet of the foxes, whose attention was at once drawn to the new comer. He caught sight of them at almost the same instant, but while the foxes were alarmed the dog was merely curious. The mutual inspection lasted for about a quarter of a minute, and then, although each of the foxes was about four times his size and weight, Tichborne dashed straight at them. They fled incontinently, and my gallant hound, knowing his own special weakness, for he never chased anything by sight that ran faster than an opossum or native cat, bustled round to pick up the scent and see these new acquaintances out of the locality.

On a Christmas Eve, three years afterwards, when coming home for the holidays, after a sojourn in Riverina, I found poor Tichborne lying dead on the very spot where we had startled those foxes. I fancied that the rabbits hopped about more boldly amongst the rocks and bushes, as though aware that an enemy that had given them many a scare was lying in the little basin below decaying in the December sun. This little fellow, although a thorough vagabond in his habits, was more affectionate than most dogs, and was never so well pleased as when allotted a seat on one of our shoulders, where he balanced himself cleverly, and looked about with a dignified "I'm-monarch-of-all-I-survey" sort of air. Even at a fast walking pace he seldom toppled from his perch. Yet he was not a trick dog in the ordinary sense of the term, and scorned the education in various clever feats that we through a keen appreciation of his intelligence wished to bestow.

His and our companion in vagabondage was a mongrel black retriever which inherited the curly silken coat of that breed, with a good deal of the shape and stature of a dwarf Newfoundland. He was named after a politician and Minister of the Crown of some celebrity both then and now. Their characters and dispositions were very much alike, so that the contemporary political history of the colony would tell all about this dog. In a secluded corner of one of the river bottoms we could always depend on starting one particular hare, that had its form there amongst the thistles. On account of its peculiarly light colour and remarkable fleetness this hare had been named Tim Whiffler, in honour of a turf celebrity, and it had run before many a crack pair of greyhounds without coming to grief. It was even recorded that on one occasion a pot-hunter from the city had turned a pack of seven greyhounds on to it without effect. When started "Tim" always went straight for the plains in the first instance, but after a few turns he, like all other hares coursed in that locality, took refuge in a deep ravine cut into the tableland, and overgrown with stunted scrub and coarse kangaroo grass, giving excellent cover. When the course commenced the politician was generally bustling about amongst the thistles, starting the rabbits from covert, and revelling in a series of short but entirely futile courses always ending in the coneys whisking their white tails in his face as they dived into their burrows.

As soon as a hare was afoot, the politician, instead of joining in what he knew to be a thoroughly hopeless affair as far as he was concerned, started off in a hurry in the opposite direction, and straight for the head of the ravine. Here he took his station, knowing that if the hare was not picked up by the greyhounds in the open it would certainly return to cover. When it did come the politician's share in the hunt was of the most meagre character. As the hare approached wrenching before the greyhounds to baffle their rushes, he made one dash at the game, but never in all our experience was that dash known to succeed. Indeed, it was of advantage to the hare, for the politician in his eagerness often got in the way of the leading greyhound, and both dogs were overturned. Melbourne coursing men, after coming miles for a private trial, were more amused by the politician's strategy in getting to that ravine before the hare than interested in the result of the course. The dog was a wonderfully systematic hunter, but after a day's beating it was necessary to place him under the shears, for his long silky hair caught and retained the burr seeds in such a way that he soon became in a sense armour-plated.

Both dogs met their death in the same way. In the lambing season flock-owners were troubled with the ravages of a horde of starved dogs, generally stray half-bred greyhounds. For these lamb destroyers poisoned baits of the most alluring kind were freely laid. It was not possible to make a distinction, and so our two dogs, whose only mission in life was that of destroying the rabbits fast becoming a plague to the landowner, were included in the common slaughter.

Silver Gums.

WHEN so many of our Australian trees were named "gums," a distinguishing prefix for each variety was clearly necessary, and so the words red, blue, yellow, white, and scarlet, as marking some particular trait in the tree, have come into everyday use. Had the pioneer bush botanist seen at least one of those trees at a certain stage in its growth, the term "silver gum" would have found expression.

In the gullies along the southern fall of the Otway Range, opening towards the ocean, grow tall blue gums, which were once believed to be peculiar to Tasmania. About the mouth of the Cumberland River, near Lorne, forests are young compared with the splendid tracts of matured timber farther westward on the watersheds of the Aire, the Barrum, and the Elliott Rivers, where the Apollo Bay Timber Company are at work. Upon these young trees the combined ravages of bush fires and a wood-drilling insect have left a singularly picturesque imprint. Some of the tree trunks are stained with the bright, carbuncle red gum itself, others are blackened and scorched by contact with the flame, still the natural vigour of the young trees asserts itself. In place of the dark green sickle-shaped leaves, which are the distinguishing botanical trait of the blue gum, a mass of young foliage glistening white in the sunlight has appeared, and when the sea-breezes pass over the valley hill-sides are a mass of waving silver. The colour fades out as the valleys deepen, and the other gums intrude their leafage until the predominant tint is once more deep green. Beside the blotched trunks of blue gums, clad in ragged bark garments, the white gums are an effective contrast. In the daylight they show a rich creamy yellow, with something of the tenderest green, rather suggested than defined. It seems more the reflection of that lusty green glowing in the foliage above than an absolute unit in the blend of colours. But how cold and white and regular these tree lines are in the moonlight when thrown out in sharp relief against the sombre background of cliff and hill—

> "Oh, moon ! the oldest shades 'mong oldest trees
> Feel palpitations when thou lookest in."

Messmate and stringy-bark trees introduce a variety of forest tones. On the most robust boughs there are spots of yellow, giving a stray touch of old-world autumn colouring to what were else a perpetual spring. These dying leaves flash here and there like acorns that ripen yellow while the oak leaf is yet dark green. After a fire the messmates are so many scorched stakes. The coir bark is so soft and inflammable that the flames dart quickly up the trunks, and for a time they stand gloomy forest monuments to all the bush loveliness that perished in the flames. Then, from root to topmost bough, green branches shoot out again, and the trunks are once more pillars of vivid green. The renewal of foliage, which in other lands is the work of the varying seasons, is here the mission of the bush fire combined with other elements.

When lightning strikes a full-sapped gum the tree is much shattered, and every vein in trunk, and root, and branch has become a channel for the diffusion of electricity. The result is a peculiar softening of the fibre, axe and saw sinking easily into what was before a tough wood. With the hard-grained, wiry, lightwood acacia quite another effect is produced. The trunk is splintered along the grain of the wood, and where the fluid passed the bark has been stripped off in shreds and strewn upon the ground beneath. The rents in the trunk are margined with greenish-grey stains where the sap flowed from the wounds. You may trace the course of the lightning upward to its first contact with an outer bough, but a keen perception would be required to find the point of contact in a red gum.

What a pall the bush fire casts over the ranges? It may be miles away, but the white vapour filters through the forest trees, hanging low over the sea horizon in long black clouds, and reddening sun and moon as they appear in turn above the forest edge. Even on the opposite range the trees are seen as in dim moonlight or through a veil. Nature puts on a robe of mystery and magnificence. These gullies seem the fitting home of legend and fairy story. When, after the fire, the fronds of tree-ferns begin once more to break and unroll in the ravines, it seems as though a long bare wall was being brightened with pictures. There will always be silver gums; for large forest reserves surely imply forest fires, and such fires are fiercest where, as along these mountain slopes, there is a predominance of acacia in the underwood. The greener it is the more resinous and inflammable. Anyone who has seen a furze hedge on fire will understand how acacia bushes feed the flames.

The silver gum country is, during the later summer at least, somewhat destitute of flowers. On the crests of the hills one finds a few yellow immortelles, with a claret-coloured orchid now and then in the fern patches. On the declines a stray twig of heath shines in the brown dulness between the trees, and at the foot of the hills are tufts of wild geranium. And yet without any apparent design in Nature, there is no lack of colour in the prospect. That which the flowers deny is supplied by the leaves and in clusters of wild berries, luminous blue, pink, and white, and the fading leaves of one of the wild geraniums lend a decisive tinge of scarlet in places. The birds that frequent the gullies are as few in number as opposite in character. From the thicket, where a couple of funereal cockatoos sweep, their long bodies aptly borne by a long, peculiar beat of wing—the hard horn bill and splashes of yellow plumage showing in contrast to the uniform black—the only other bird song is the twitter of blue wrens. Down among the roots of a peppermint bush, sheltered by creeping plants, is the wren's home—snuggest and neatest of all the pear-shaped nests. The little ones inside may just hear the impressive roar of the far-off sea, barely loud enough to drown the twitter of the old birds as they come glancing along from an insect hunt amongst the wild briars, their long shaft tails catching the sun rays as they dart amongst the brambles. Even the habits of Otway cockatoos vary strikingly. Along the coast the black ones fly in pairs; but on the northern verge, where lightwoods take the place of the silver gums, the white sulphur-crests are in hundreds. They are loyal birds, very constant in their affection for a particular stretch of forest, and have held joint possession here with white men for nearly half a century. As they fly across the road in the strong light of noon, that delicate flush of saffron which burns in the under-wing shows very beautifully in the semi-transparent feathers. Although few birds are seen amongst the blue gums, large red and black butterflies, and others whose black wings are barred with white spots, frequent the sides of the ridges, their eccentric movements making kaleidoscopic effects in colour.

On the boughs one finds, too, a rare type of mantis—green like the ordinary insect when young, but with its colour fading in old age, so that, when fully grown, it resembles closely a dry twig. The body is irregular in shape, just as a small branch is uneven in outline. The legs join the body, not in uniform pairs, but singly, as leaves connect with the bough. Although the insect when full grown is some four inches in length, it cannot, under ordinary observation, be distinguished from the bough except when in motion—a striking illustration of Nature's method of preservation based on the law of harmony.

The scarcity of pasture grasses and of edible wild plants or vegetables is a noticeable feature in the blue gum gullies. The root of the ordinary bracken fern, or the white juicy heart of a tuft of grass tree—which is full of a starchy sweetness—may both in serious straits be baked and eaten, but neither is quite so good as the yam or sweet potato of the South Sea islander.

A variety of mustard growing along some of the mountain creeks is the one salad amongst the marsh plants. How the myriads of kangaroo, which the early Otway pioneers found everywhere along the coast, picked a living from these sparse grasses is a matter for wonder. Few of the native tribes ever came so far south unless coasting from either wing of the forest, and hence the marsupials multiplied without check as the rabbits do to-day. Black wallaby are thick enough, even now, amongst the coast scrub. The name varied with the aboriginal dialects of this portion of the continent, being "kurün" in two of the native tongues, and "kurræ" in the third.

According to the traditions of the bush—not always reliable—the name of kangaroo was given under a misconception. An aborigine being asked by one of the early discoverers the name of the animal, replied, "Kangaroo" ("I don't know"), and in this confession of ignorance or misapprehension the name originated. It seems absurd to suppose that any black hunter was really ignorant of the name of an animal which once represented the national wealth of Australians as the merino does to-day. The skin of the big kangaroo was the coarser robe which the western native, like the Russian, wore with the fur inward. The skin of the black wallaby made a water-bag, used very much in the same fashion as the goat-skin of the Eastern water carrier. The pouch served as a water-bucket too, and the strong sinews were the natives' thread. The long shank-bone, when pointed and spun between the hands in a wooden slot, was their fire maker.

The animal had no place in native superstitions. Their evil spirits were generally such living things as are unfit for food. The native heaven was a hunting-ground so thickly stocked with kangaroo that not even a bush screen was required to stalk them.

At present the largest game in the South Otway forests are the wild cattle. The herds had their origin in the strays from the old cattle runs beyond Apollo Bay. The cattle were driven through the forest to the Geelong markets years ago, and as the tracks were narrow and scrub-bordered, a large percentage were lost. The moose and elk in Canadian pine forests, or the stag amongst the Highland corries, is not harder to approach and bring down than these "clear skins" or unbranded wild cattle. You may ride for days through the scrub without finding their tracks, and then, unless the approach is warily made, the herd is certain to take alarm. Rushing from their feeding grounds to some well-known trail, they trot away for miles, and finally cluster together in the heart of the densest scrub.

In order to understand the cunning with which wild cattle hide one must have sought for the calf a few days old that an ordinary station cow plants away so carefully. How faithfully the little fellow obeys the instructions, given in some way or other, to keep to his hiding-place. You may almost step over him in his tuft of grass, but he stirs not. Pull him from his nest, and he bellows so lustily for help that the old cow comes tearing in from the plains, ready to annihilate all intruders.

In-breeding makes most of the wild cattle thin and ragged. Some of the cows are, however, square-framed, and the old bulls are magnificent, deep-chested fellows, shaggy as buffaloes.

These wild cattle are thickly covered with long dark hair, something like the coat of the West Highlanders. You may trace in some of them the points of the old-fashioned breeds ; but their wintry life in the ranges and gullies has developed, or revived, distinct characteristics. In moving through the scrub or along their trails, which wind about the bases of the hills in true cattle fashion, the older bulls take the lead. From the habit of using their heads as a battering ram their horns are almost cut through to the forehead.

The tough cord-weed, found everywhere amongst the tangle of the Otway Forest, is barbed like a fretwork saw, and this is the chief cause of the havoc wrought on the wild bulls' horns. Sometimes hunting parties are out for a week before getting a glimpse of a herd, though they may have passed within a hundred yards of scores of wild cattle. Unless the country is thinly timbered the hunter, as soon as he sights his game, leaves his horse and follows on foot, being able thus to move faster than the cattle, unless they strike a favourite trail.

Wherever the Australian coast line is particularly bold a faint likeness to familiar objects may be found either in obtruding masses of rock or the indentations between. Such are Lady Macquarie's Chair, overlooking Sydney harbour, the organ pipes and the head of the Iron Duke upon Mount Wellington; while many landmarks familiar to mariners bear some resemblance to ordinary animate forms. The strange-looking Sentinel Rock, a recent one of many quaint designs wrought in chance fashion by the sculptor Sea along the southern coast of Victoria, is a wild-looking head that peers out and above the Artillery Rocks, some eight miles westward from Lorne, and, fortunately, rather beyond the tourist's range. From the ground to the crown of the head the height is twenty-four feet, and the bunch of coast scrub, which has found sufficient sterile soil in a cleft of the rock to give it a dwarfed existence, looks like the last ragged tufts of hair clinging to the weather-beaten scalp of this coast wizard. The guns underneath are strangely realistic too; and, curiously enough, the same rocky table that carries these stone cannons shows, partly embedded, spherical blocks of sandstone that might almost serve in case of emergency as projectiles for the old regulation smooth-bore artillery. What traditions of the Armada and of threatened invasions would have been associated with that weird head had it looked sea-ward from England's cliffs instead of being on sentry here by the peaceful southern ocean! It might have typified the natural island spirit, watching jealously, untiringly, for the foes that the political convulsions of the continent close by perpetually threatened to hurl against it.

The Otway Ranges rest upon a great bed of sandstone. Following up the creeks, which trickle down between each mountain ridge, the sandstone shows in beautiful natural stairways. Near the mouth of the Cumberland some convulsion of under earth has tossed it into towering pillars and rugged cliffs. Round about sea eagles wheel and scream. From these high points the telegraph line to the Otway takes a long drooping fall to the hill-slope beyond. Some bands of the sandstone are softer than others, and here the sea cuts out quaint fissures and caverns in the cliffs. Within a couple of miles of the Sentinel Rock the Cumberland Cave, though small as compared with some of the mysterious sea chambers farther down the coast, is yet an object of interest. The stalactites which adorn the roof and walls have been broken away by desecrating tourists, who would not pillage a church or a shrine, but rob nature's sanctuaries without a single twinge of conscience. The floor of the cave, only reached at low tide, is loosely paved with small water-polished pebbles. One of the few pendants yet remaining in the cave is of oval shape, and not unlike a large soda-water bottle. Beyond Cape Otway the ravages of the sea are more startling in their nature, and sea caverns penetrate far under land. One of these was the tomb of some seventy unfortunates, who sank with the ship *Loch Ard*. Out to sea, where the cliff once stretched, limestone columns stand up like sea monuments to the memory of those whom the ocean destroyed. The ship's yards rasped warningly against one of those outposts before she struck that wedge-shaped island rock and sank in its shadow. Here, hard flakes in the cliff jut out, and stand isolated when the softer stone has been scrubbed away. Under these ledges swallows build. Upon the balconies the older birds watch the long seas rolling in to them, and yet the wildest wave cannot fling its foam high enough to damp the brown mud nest or wet the downy backs of the fledglings clustering together. The homely and the impressive, the beautiful and terrible are neighbours here. Between the headlands are bits of shingle strewn with white shells. On calm nights I have sat upon the ledges, and, as the billows curled over and broke, watched the phosphorescence roll upon the sand in long curves of yellow light. The fairy laboratory of geology is amongst the minute and beautiful forms of carbonate of lime which we find imprisoned in snow-white cells in the heart of a block of bluestone, but for giant's handiwork one must seek in that studio where the toiling, persistent sea has been busy for ages.

Although for the last ten years Lorne has been one of the most popular of Victorian holiday resorts, it keeps its freshness wonderfully. The range of mountains that curve round like a shepherd's crook, and cling about the little sea town, are so impenetrable that one need only break away a few paces from one of the blazed bush tracks to feel that he is wandering in a virgin forest, as silent and wild as the first explorers of these coast mountains found it twenty years ago. One may even experience that novel sensation of being utterly lost, while he can yet hear the perpetual "break, break, break" of the surf upon the yellow sands, and can see dimly through the green curtain about him the blending of sky and sea. It is this happy association of ocean and forest that gives the place its charm. The odour of the sea is tossed in from the south, the scents of the woods fall softly down from the hills. To-day you may wander about the shore studying marine life in those crystal miniature aquariums left by the receding tide amongst the rocks, or may watch the amber billows curl over until the patches of broken water stretch out into one white wall that in a second dashes upon the shore, and fritters away at your feet in flounces of snowy foam. To-morrow you are up in the ranges, picking out natural vignettes among the gullies, and wishing in a patriotic way that some capable artist could see and reproduce these specimens of Australian scenery.

Lorne is a land of waterfalls. It may be truly said of it—

> " Day and night to the billow the fountain calls ;
> Down shower the gambolling waterfalls
>   From wandering over the lea.
> Out of the live green heart of the dells
> They freshen the silvery crimson shells,
> And thick with white bells the clover hill swells
>   High over the full-toned sea."

The finest of the cascades along this southern mountain slope is the Erskine Fall, which rolls gently over the crags, and falls its 130 feet in a varied shower rather than leaping out in a volume to the pool beneath. Tennyson wrote of the "waving tapes of waterfall." Looking at this fall one realises the absolute perfection of the comparison, for the Erskine is a mass of silvery tapes and streamers created by the jutting points of the broken wall down which it filters. The Horseshoe Fall, the most remote and rarely visited, was so named because of the reversed arch of granite through which it pours. The view to the southward from the cliff takes in a wide and charming landscape, for the valley opens out towards the sea. At the edge of the pool, drenched with the spray of the cascade and falling down with the valley, is a ragged fringe of mossy carpet covering the floor of a very hall of ferns. The wide-spreading crowns of the tree ferns make a roof through which only quivering scraps of sunlight fall, and the tall trunks, clothed on one side with a red rusty moss that gives an extra tint to the colouring of the glen, are the supporting columns that complete the natural architecture of the place. At the Phantom Falls the water comes down in a double leap, first to a little stone platform midway, and then with a final spring into the pool beneath. The white foamy water is set off with a background of dark green moss covering thickly the bank behind, and flaunting in long tresses, so that if the water suddenly dried up, the appearance of a cascade in green would still remain. Where the water strikes the stone ledge beneath hundreds of little sparkling fountains arise like heavy rain-drops splashing up from the pavement in the lamplight of our streets. Long ago some pioneer tourist announced the discovery of this new waterfall, but as nobody else could find the place it gained the name of Phantom Fall. At the foot of every cascade one finds the same picturesque-looking pool, darker than usual, when the waters have filtered from the uplands, where forest fires have left nothing but blackness and desolation. The burned forest is popularly, but wrongly, depicted as the dead forest. In the fire country the leaves have whitened after the flames passed, and they cling to the branches for a season until superseded by a fresh growth. The big forest trees are only wounded, not destroyed. But the forest "bark ringed," or killed, in the interests of husbandry is an assemblage of whitened skeletons. Most artists either cannot or will not paint live gum trees—common as they are in every bit of Australian bush. But fewer still attempt to reproduce for us the weird look of the dead gums. Upon the mountain-top they come out hard and sharp against the blue, with no softening of the edges, no blending of colours.

In speaking of the list of show places it would be unpardonable to omit mention of the beauties of Paradise and Cora Lynn; the latter, in an irregular way, a repetition of some of the buried New Zealand terraces, done in brown sandstone. From step to step of a giant staircase the water leaps in a succession of cascades, and the music of these forest fountains tinkles in the faintest of liquid whispers up the sides of the quiet glens. This Paradise, over which the grand old gums stand sentinel, has gates of flint— not of jasper—and the "narrow road" by which the mountain pilgrims come is festooned with ferns and aromatic shrubs, and blocked and broken with dead tree trunks and rugged boulders. The conventional heaven has been built out of old-world materials, as the paradise that Mirza saw in his vision was Oriental. A paradise to satisfy Australians must have some of the elements of beauty that throw a charm over this wooded glen of the Otway Mountains. The heaven of city people is perhaps an architectural dream—a vision of towers, faint and far away like Camelot; but to those who love the country—their own country—and its traditions their heaven to order must have some of the sylvan charms of this neighbourhood which I am describing. It irritates our conceit to think that scriptural imagery in this matter was limited so closely to that little bit of Palestine, and that in biblical dreams there was no vision of the "glorious marvels of the years to be"—no suggestion of continents washed by western and southern seas.

To those who have good legs, and for whom loneliness has charms, one of the chief attractions of a Lorne holiday is the long walk up the valley of the Erskine River—over the piled boulders and tree trunks, through avenues of fern trees, and into secluded bits of forest where for half a century the sunlight has never penetrated. The thousands of trees hurled by winter tempests from the heights above lie about, wrapped in mossy vestments of pallid green and bronze. Where the rocks are bare hundreds of pretty brown lizards are basking, and the larger, but less handsome, blue-tongued lizard rustles through the scrub near the edge of the stream. It is interesting to watch one of those graceful creatures catching insects. Occasionally a flying ant pitches upon a neighbouring rock, and the lizard seems to mark him down just as a sportsman would his game. Slipping quietly from his look-out, the lizard glides about the base of the rocks, and in a few minutes his little snake-like head, with a pliant black tongue in active anticipatory motion, comes stealthily over the ledge close to the musing ant. As a rule, the insect seems to have expected such a visitation, and leaves at once in search of a safer resting-place. In lizard logic it seems to be an accepted fact that where one fly has perched another may come, so the little hunter in his suit of shining bronze settles down in the new position.

On the stone terraces, where the water comes rushing down in miniature rapids, it is interesting to watch the tiny stream trout when surprised in the sunny shallow pools rush up in the very teeth of the current for perhaps a dozen feet, to gain the shelter of the home pool above. A wavy line in the water marks his course, and in turning the little fellow throws himself to one side, where he is held by the streaming moss until ready for a fresh effort. It is a marvellous feat for these little atoms of silver and green to accomplish, and, looking at it, one understands the salmon's progress up the mill weirs on English rivers. The appearance of an intruder close to the edge of the large pools startles scores of the larger brown stream trout, who flounder lazily away into the darker water or the shadows of the boulders, but soon return again to the current to wait for any food that may come floating down with the stream. In the dusk of evening the surface is in commotion as they rise to feed, and if these Erskine fish are not the descendants of English trout their habits very closely resemble those of our fine freckled friends of Riddell's Creek and the Watts. One of these dusky-skinned fish flaps awkwardly on his side in the current as though injured, but shoots away as gracefully as his fellows when alarmed. It subsequently transpires that he has lost the sight of the left eye, and is very shrewdly making the most of the remaining one. This fish has adapted himself to altered circumstances. There is very little bird life along the stream, but now and then a bit of gay plumage flashes out for an instant. Here it is a pair of lory parrots in a silver wattle, and tree and birds combine to make a little picture not easily forgotten. Lories are of the deepest crimson, two patches of dull azure marking the wings, and the white beak contrasting with their gay costumes. The silver wattles are worthy of the name, the young saplings, veritable rods of silver, being spangled more beautifully even than the cedars in early spring, and as never holly or laurustinas were decked by winter frost. The little redbreast seems to find a congenial summer home in this retreat, but his plumage is less ruddy than when we saw him about the city fields last winter. Now it is something of a rich orange, and more nearly approaches the bright buff of the English robin. The deep silence of the gully is broken now and again by the scream of the black cockatoo, the most untamable of Australian birds. Near the pathway leading to Cora Lynn two fine trees have been barked to a height of 150 feet, and are conspicuous items in the landscape. Here the boulders are dark and curiously water-worn, but a few paces farther on in the shadow they are rocks no longer, but huge masses of something wrapped in the soft mosses of centuries. All is fern world, and one cannot tell where the ferns end and the mosses begin. There are really no breaks in the vegetable chain. The links of relationship are about somewhere if we but take the trouble to seek them out. The great tree fern is the king of these gullies and the giant of his race, and it is a long gradation down past the batswing with its tasselled drapery to the glen fairy—the little delicate maidenhair, hiding modestly in some sylvan bower. It loves the secluded cleft of a bank or the shelter of a gum bough that has been wrenched from the trees above, and has whitened for a season down in the shadows. The damp logs are ferneries of themselves, and one can pick many tiny specimens, some with beautifully-marked fronds, from the lawn of moss.

The close atmosphere of these lone gullies is suffused with the scent of musk, intermingled occasionally with just a suggestion of fragrance from the white-flowering gums, which, when fully clothed with blossoms, are beautiful landmarks along the slope of distant ranges. It is very pleasant to rest after one's climbing in such arbours. The dingoes, stronger and fiercer here than in any other portion of the continent, are dying out, and in their haunts English foxes establish themselves. The old order changeth, giving place to the new.

The first glimpse of blue sea through a forest trellis of green and yellow is worth the whole journey, and from this altitude it looks vividly blue. In different places along this ocean slope we find large patches of young wattle scrub springing luxuriantly, and may conclude from this that within the last few years a bush fire has swept across this part of the country. Wattle scrub is brought into existence by influences just the opposite of those that induce the growth of other young trees. The seeds of most of our forest trees germinate when a sufficiency of moisture is provided; but the black seeds of the wattle, which retain their power for years, are preserved in the mould without being in any way affected by moisture. The fierce heat of a bush fire cracks their hard outer casing, however, and in the following spring young trees cover the land thickly, where, perhaps, for half a century the air has never been sweetened with the fragrance of a wattle blossom.

A night in these haunts of Nature is singularly impressive, whether it be in the intense quiet, when the solemn chant of the mopoke alternates with the eternal inward rush of the surf, or when a storm is raging amongst the hills. A storm up among the ranges is a majestic function of Nature; but to realise it one must have stood upon one of these high Otway peaks at night, and discovered himself an intruder in the haunts of the lightning. Down in the valley electric streaks shoot along above the trees, burnishing the stream, and searching out every detail in the forest. Then everything is blackness. In the morning a white cloud of mist fills the valley, and as a gust of wind breaks round a corner of the hills the shroud is lifted for an instant, and one has a glimpse, far away down, of a settler's hut, with the smoke clinging about the chimney. Then the white ghostly curtain settles down once more, and all is loneliness and gloom. These storms, when succeeded by instant sunshine, are the distilleries of Nature, and, by the power of a chemistry that we can only vaguely comprehend, perfumes come out of the drying woods that are the real elixir of life. Under their subtle influence even the most languid and morbid of mortals is forced to the realisation of the intense happiness of existence.

The sunsets seen across these mountain tops are some of them very beautiful. The peak behind which the sun is dipping seems wrapped about with a yellow haze, and as the sun disappears a mystic avalanche of light rolls into the valley at your feet and fades away. In one instant you have the finest possible illustration of the splendour of sunlight, and in the next, from mountain top to mountain top is nothing but shadow.

The coach ride across the mountains is an agreeable incident in a Lorne holiday. The winding track leads over the highest points of the ranges, and some fine distances open out on either side, though from foreground to horizon the eye finds no resting-place in the wide area of tree-tops. The harvest is here ripe, and waiting only for the splinter and the locomotive. Ere long the axe will ring and the whistle echo amongst the hills; but it is comforting to know that for a century to come these sounds will not break the solitude of the gullies on the ocean side of the range. Down in the foothills the lightwoods give a park-like aspect to the country, and some of the rushy pools by the road-side are beautifully starred with water-lilies.

Village and Farm.

My village is set deep in a hollow of the plain, so that you almost stumble into it over the hill-tops capped with grey basalt. These hills seem like a barrier shutting it in from the rest of the world. From the table-land above you can see the black clouds of smoke rising above the city, and the masts of shipying in port, each offering its suggestions of busy commerce. You may fancy that you hear the din of the Babylonian chorus—or that you feel the throb of a thousand engines and the vibration of a thousand cranks, where the tall factory chimney stacks throw off their contributions to the dark canopy. But down in the valley beneath there is no re-echo from the city. It is a peaceful place.

The houses are not glaringly new as in those mushroom towns of the north, but they are homely and comfortable. The white paint, long ago faded to dull drab, and half hidden beneath creeping ivy, clinging honeysuckle, and sprays of intrusive passion flower—is a pleasant contrast to the glare of new pine boards. In every corner and hollow of the valley trees have long since been planted—here an English oak, there a cedar of Lebanon, next a Scotch fir, and farther on a black Austrian pine—a vegetable community as cosmopolitan as the people of the village, who are made up of many nations—Englishmen, with an abiding belief in their own land, and a faculty for copying its traditions and institutions; Irishmen, with that keen love of country that has wrought so many misunderstandings abroad and heartburnings at home; Scotchmen, thrifty and rugged, like the shelties and black cattle of their Highland hills, gloomily prophesying a future of sorrow and disaster for this new land because the village boys play at cricket on Sabbath afternoons in one of the bends by the river side. This British composite is leavened by units of other lands, who have almost forgotten their nationality.

There are strange colonial experiences here. Some

of the villagers were the gold diggers of thirty years ago—men who burrowed for wealth beneath the white hills of old Bendigo, when the city was of canvas, with no bright green elms lining its quartz roadways—men who tell tales of these old-deserted claims along the scrub mounds, as dramatic and as inexact as Joaquin Miller's stories of the rugged Californian gold-seekers of "'49."

In the centre of the valley there is a large pond, with an island in the centre, and all along the margin a wealth of pendent pale-green willows. Years ago the pond was a little lake, the home of flocks of water birds, and girt about with stately gum trees. But when the settlers came and tore up the green slopes with plough and harrow, the storm water from the hills brought down the surface mould, and so the lake was silted up. Less than a half-century ago sedate emus trooped in stately columns over this hill-top, upon which the figure of a hare on its way to the orchards is now outlined in relief against the flush of dying twilight in the west. Kangaroo came out into the moonlight from the hollows. Now the white tails of many rabbits twinkle in the dusk. Where the highly civilised geese are nesting under the willow fringe, swans laid their long white eggs. In the little cottage gardens everything is old-fashioned. The borders are of thrift or rosemary, and the fences are hidden in thickets of golden broom or pink-flecked sweet brier, filling the little garden with its fragrance. Chrysanthemums of all shades spread their glory over the flower-beds in the autumn. A white trumpet-lily has taken absolute possession of one corner, and close by there is a huge lilac bush crowned with blossoms. Out in the fields the ugliness of post and rail is sometimes hidden in a dark green covering of furze, or a square-cut hedge of hawthorn, along which the children run on Sunday afternoons, searching for a rare spray of pink in the ridge of snowy almond-scented blossoms. In the kitchen gardens huge elderberry bushes hang their flakes of white flowers. Before the sparrows came, bunches of dull, claret-coloured fruit followed the blossoms, but now they never appear.

Can anything be old in this new country? From an antiquary's point of view, perhaps not. But a lifetime is a long time. No memories can be older, and it is memory that makes this little old-fashioned village the dearest in the land to many who have gone out beyond its limits but not its influence. In that little circular cemetery, for instance, down beneath the long brown grass that waves in the summer wind, there are memories as dear as life-blood. Every thought here is a gem precious beyond price, and memory, like a rich and thrifty matron, pours out before us all the wealth of the vanished years—nothing squandered, nothing lost. No Lethean draught to drown such sweet company, rather the fruit of the enchanted lotus stem, that we may

> "Muse and brood, and live again in memory
> With those old faces of our infancy,
> Heaped over with a mound of grass,
> Two handfuls of white dust shut in an urn of brass."

And when several generations have lived beneath the same roof-tree, and some have died beneath it, every shingle is sacred. Every old-fashioned flower in the garden has associations to cherish, whether the blush roses climbing about the windows or the tufts of white and golden guelder rose nodding in the breeze.

If the village gossips in their hours of idleness are sometimes busy with other reputations than their own, there is gold down beneath the plain earth. The little community has a noble heart that throbs with a misfortune made universal, or is rent by a common sorrow. Some morning there is a hush in the wide grassy streets, and the children no longer clatter and laugh along the gravelled pathways. Death has come in the night, and although sympathetic sentinels are on guard, has taken one spirit away. The seal of sleep eternal is on a white face that will never brown again in the healthful sunshine. The darkened room—a symbol of abiding grief within, and without a sincere sorrow—is sweetened with flowers. No floral gem is a gift too rare for that cold white casement of a soul. So the neighbours gather their heartsease with the dew-tears fresh upon it, and white double stocks, and wallflowers and blue forget-me-nots to send as a last offering. Perhaps one of the grey fathers of the hamlet has passed to his rest. Then the village clergyman leaves for a time the tiny human speculations on Infinity that pass for sermons, and preaches the gospel of condolence and affection in words that soothe heart-sickness, like the sympathetic whisper of woods and waterfalls.

The political centre of the village is the blacksmith's bench. If there is a chance idler, he comes here to pore over the newspaper that by noonday is always black with the smoke and cinders from the forge, and marked with the thumb-prints of many readers.

Farm hands with horses to be shod, the crank of a mowing machine to weld, or a ploughshare to be "set and steeled," drop in, and having exhausted topics of local interest, such as the price of hay, the qualities of a certain strain of draught stock, or the probable harvest yields, they relapse into politics. Broad questions, such as protection, free trade, or secular education, are the subject-matter for argument. As there is not an acre unalienated for fifty miles around land acts have only an antiquarian interest. The subtleties of lobby politics or corner complications that so interest the journalist, and are gossip for the city man, rarely penetrate to Arcadia.

The older people are conservative both in habit and opinion, while the younger, like most colonials, are deeply imbued with the spirit of a new democracy. They have no traditions to cherish, no institutions beyond those of State and Church to maintain. Self-interest is the secret of their concern for one, and they are loyal to the other from mere force of habit. Their fathers made the Church an important institution in the land by the power of prejudice and party feeling. It was woven into their politics, so that the two could not be dissociated. But the old spirit of intolerance and bigotry that built up mountains of rancour washed about with seas of blood is a sentiment with no meaning for their children. Without emotion or regrets they see the old denominational differences dead or dying about them, hard and forbidding to the last. They only say as Dickens said with such a different meaning, " Dead, right reverends and wrong reverends of every order, and dying thus around us every day."

The social pleasures of the village are few. In

music the concertina was once a potent power for melody or torture just as the tastes of the listeners inclined. Then there was a real Irish fiddler who tore madly through "The Blackberry Blossom," a quick step that the operator spoke of endearingly as "a horrible fine chune." One of the Scotch villagers, too, was the owner of a well-worn pibroch, and sometimes at night the echoes of the hills were busy with the quaint Gaelic music.

There is little variation in the method of farming. When the fields are weary with the giving of their strength to so many harvests, they can rest for a season. There is no mortgage on the farm, no lien on next year's crop to draw every possible corn blade from the soil, and exhaust both home and husbandman. There may be little wealth, but there is no poverty. No home-sick Ruth has to glean in the cornfields. Indeed, the Australian Ruth either drives a pony phaeton, or is at least the charming autocrat alike of parlour and dairy. And on a hot day in midsummer what sitting-room in the land is so pleasant and wholesome as a clean, cool country dairy?

Of course the village possesses a tragedy. On a Christmas Eve, years ago, two farm-labourers were at work in an out-of-the-way field, and the old tragedy of murder, with avarice as the motive, was re-enacted. An old man was chopped down by an ex-soldier with a hoe, as one would fell a thistle in the corn, and the body was sunk in a narrow bend of the brook under a shelving bank, where it was hidden from the daylight. Boys came as usual and lounged on the green bank, while they whipped the pool for minnow, but never suspected that such a ghastly remnant of mortality was lying beneath the ledge. The village policeman was something of a philosopher, and believed in the sensitiveness of even a murderer's conscience. He had his own theory about the mystery of the old man's disappearance. Pretending to believe that the missing man was drowned while drinking at the stream, he asked the suspected man to point him out all deep pools in the brook. The first one that he showed any inclination to avoid was the only one searched. Then, when the body was found, and the fact of a murder proved by a score of cuts and bruises, evidence poured from every quarter. Those strange impulses that repeatedly misdirect erring humanity had prompted this criminal to do a score of things that brought about his own conviction and execution.

The village has a periodical awakening at harvest time. For a few days scythes have been busy mowing roads through the fields of oats, the tops of which show just a faint tinge of yellow—the first sign of ripening. Then the din of mowing-machines is heard from every field, and the rattle of hay-rakes, as the curved iron teeth collect every scattered blade, make a true harvest chorus. In the orchard the cherries redden and ripen; but the glowing clusters must nestle undisturbed amongst the dark burnished leaves, until the fields above show nothing but stretches of short stubble. Then the fruits are gathered for market.

The hay harvest anticipates the grain, and hundreds of the drifting population of the land — men who work in cycles—move a stage farther up towards the tropics to assist at the garnering in the great northern wheat fields. In the early summer they are still farther north, shearing the sheep in a Riverina wool-shed. Then lower down, where the season and the clip are more tardy, until a southern midsummer finds them back again in the hay fields. They move on a ceaseless labour circuit, and with the ebb and flow of this tide of humanity a new feature in the life of the village is annually born, and just as surely fades out again with the dying year.

Once the wide plains to the south and west were one vast common, or grazing ground, the pride of the farmers and householders, and the envy of the large grazier, who had no privileges of pasture there. Every morning mobs of milking cows, followed by shouting boys and laughing girls, came trooping down over the hills to the milking-yards. All the grassy stretches of upland were deeply seamed with a hundred home paths made by the cattle tramping in each other's tracks year after year. When cattle travel continually on a narrow road they cut it into regular steps or ridges crossing the line of march, and resembling the furrows in a ploughed field. But on the open plain they form parallel lines, and follow the beaten tracks.

The old cattle trails can still be traced through the scores of paddocks that were once the plains. Some were like the railway lines running out from a capital. At first the main trunks, broad and distinct; then the branch tracks striking off, until finally lost in some favourite stretch of pasture land. The trails told the habits of the cattle. They shirked the high ground until the last, winding away around the cliffs and the bases of the hills. In the valley the trail was broad and brown, but on the hill-side it spread out into a great fan pattern.

The herdsman had a double duty—to prevent his own stock from wandering, and to see that the flocks of sheep and the mobs of cattle market-bound did not linger on the plains longer than the law permitted. There was the annual branding time, when thousands of cattle were mustered, and as they rushed through a crush passage, a big tar brand—the initial letter of the shire—was pressed on their glossy, throbbing flanks. It brought a part of the coat away, and the new patch was always of another colour, so that a white cow had a large red letter on her flank, a brindle came out black, and red hides were initialled in dark plum colour. Once a year the cattle cast their coats, and the branding was repeated. In time a wave of "selection" swept across the plains, and the common was reduced to a few square miles on the verge of the village. Farmers were shut out, and the grazing privileges limited to householders with no acres of their own; and as the reserve grew smaller the contest for possession was more keen. Every form of mean piracy was tried. Dealers from the Melbourne cattle-yards, with surplus flocks on their hands, sent them out over the open country on the pretext of "travelling." They even rented paddocks on either side of the plain, and fed backward and forward across it. Finally it came to a declaration of war between the villagers on one side and a few stock dealers with a long retinue of cattle-yard loafers on the other.

Hostilities commenced one Sunday afternoon. A group of village youths were on sentry in the hills when a force of men drove up, each armed with a pick handle or a loaded whip, and the sheep were turned out as a direct invitation to battle. The acceptance of the challenge was prompt and decisive, The boys, who have ever been noted as fast bowlers in the cricket field, would not come to quarters, but engaged at long range. There were round cobble stones like cricket balls all about the hills, and with these they swept down upon the invaders in a hailstorm of round-arm bowling. It was a band of wiry, active Cossacks, pestering a regiment of heavy cuirassiers, and the biggest were beaten. Next the sheep-owners and shepherds came each armed with a gun, and sat beside their flocks; but the defenders, making a demonstration in force, dismayed and disarmed them.

A white track winding up amongst white-trunked trees to where the old farm sits in the saddle of the range. The slope was once a length of bush highland, but is now a park. The gum trees, thinned in numbers, have broadened in shape, each throwing its arms outwards, as though seeking always for that touch of companionship lost nearly fifty years ago, when the saw-millers passed through and cut away the straight trees for timber. The denser thickets, once the home of the wild pigeon and the lyre-bird, have been cleared away, and only a strong lightwood or a wattle nestles here and there between the taller forest trees. On the opposite slope a long garden stretches down to the creek—a garden with poplars towering in the corners, and broad paths margined with red currant bushes and miniature hedges of thorned gooseberries. In the large beds the nectarines, blossoming like the magnolia without sign of leaf, are robed in rose pink. A breath of wallflower breaks from the old-fashioned garden-haunt of peace and ease, and carries its perfumed memories to the swagman tramping along the dusty country road. At the foot of the garden there is a pool where clumps of lance-leaved lilies droop over the water, looking beautifully white and distinct when seen in the moonlight. Sometimes under the shadow of the leaves an eel curls the water into an eddy with its tail, but unlike those of the Botanic Garden ponds, which come to the surface in broad daylight and break bread with the civilised goldfish, the rustic eels are shy. Even to them the universal peace of the land should be some assurance of safety. In the darkness the one sound from the pool is the regular metallic clank, clank of a night-frog, the note resembling, more than any other sound of the bush, the distant beat of a bullock-bell. Along the bank a handful of brown mould trodden fine as flour trickles down amongst the lilies, showing that a rabbit or a water-rat has passed along. In the cool of summer nights the workers of the farm come down here, and, lolling on the green bank, fish for eels, with wattle rods strong enough almost to stop the rush of a white porpoise. Some of these toiling Ishmaels have fished in many waters. They have tethered their spare codfish on the banks of the Murrumbidgee while waiting for the shearing; have caught sea salmon in the Gipps Land lakes while returning from some false alarm of gold on a far-away tributary of the Snowy River, or drawn fat and yellow perch from an ana-branch of the Murray, while the wheat was ripening.

Away out amongst the wattles a herd of white-faced cattle are grazing, the rich red of their flanks curling into little ripples of hair. In a new bush country the cattle have their necks blackened by friction with the burned stumps, but here they are spotless. Scattered about among the trees are fawn and silver-grey dairy cattle, with sharp black chines and ebonite horns curling up towards each other in perfect crescent form, and tipped with a tinge of luminous soapy yellow. One of the mothers of the herd is dreamily chewing a bleached bone she has found in the grass, and in this simple act telling an observant pastoralist just as surely as any chemical analysis could that there is a want of natural salts in his pastures and of lime in the waters of the stream trickling at the foot of the hill. These browsing cattle are as ornamental as deer in an English park. Some miles away there is a little circular hill where, years ago, some imported deer were set at liberty; perhaps once a year a glimpse of their red flanks is caught as they dart away amongst the sheoaks. From the topmost pinnacle of this wooded hill there is a long view, extending in summer over a sea of whitened grass, with an arm of timber curling round in the distance, like the outline of some new continent. It is a perfectly placid sea—a contrast to those billowy Wimmera plains, known for so many years as the "Bay of Biscay," and where, according to popular theory, the burrowing swamp cricket has for league upon league broken a naturally level plain into mounds and ridges. Everywhere about this farm are suggestions of the old world which one never sees in the "selection" areas of the wheat country to the north. On the banks of the creek the blackberries have grown into a great tangle of brambles. How many centuries would be required to gain this luxuriance in their native woods? Still more pleasant will be the reminiscences of motherland when the thrush, the blackbird, and starling are quite at home in this long garden. They are coming surely if slowly. In a single week I have seen two blackbirds not so many miles from here, and the thrushes are thick beneath the trees. These home birds are a pleasant surprise to most of us when seen for the first time. We feel much as a long-absent Australian would feel if, in Hyde Park or the Bois de Boulogne, he heard in the gloaming the clear rolling note of a laughing jackass amongst the trees. The first blackbird brought me to "attention" in an instant as he darted across the path from a stone pine speckled with little honeycomb plumes. There could be no mistake as to his identity when he flashed in the sunlight, chased away by a brown thrush, jealous, perhaps, of the presence of so interesting a rival. The yellow bill shone beautifully clear in contrast to the velvet-black plumage. A starling's bill would have been fainter in colour, and the green spangles in his coat would have shone more brilliantly, but less blue-black, than in the case of the blackbird. A momentary glimpse of the starling suggests the green in the neck of a mallard drake, while the two colours of the blackbird remind us of one of the satin bower birds. In the study of both there is a fascination for which either the link of relationship or Richard Jefferies may be responsible. Whichever it be, most of us will say, "Prosper, thrush, blackbird, and starling; knit the links of sentiment closer still with your sweet song and sunny presence."

J

In this farm garden as yet only lory parrots show their crimson in the trees, and yellow-legged minahs, tamest of all Australian birds, fly into the big farm kitchen—where even the chickens dare not venture, and pick crumbs from the earthen floor. Looking abroad through the pastures it is noticeable that while the wattles and gum scrub spring freely on every patch of unmolested land, neither the deciduous English trees nor the pleasant-tasting sheoaks spread beyond the limits of the plantation. The cattle see to it that none of these attain maturity without protection, but they touch very few of the native trees. Wherever an Australian forest has been cut away it will renew itself in time if the surface is unbroken and the paddock not overstocked. All about the bases of the dead stumps the crust of earth is forced upwards, as though mushrooms were breaking through. This is caused by the catacombs of the sugar ants. In these winding tunnels are stored the sweet white bread gathered beneath the manna gums, and one may inspect the storehouse without fear, for the sugar ants are a peaceable community.

Old farms are always rich in picturesque elements. In the marginal bands of the field—that unploughed space between the fence and the nearest furrow—there is a self-established community of flowers as cosmopolitan almost as the occupants of a modern garden. Year after year they have come here with the seed of successive crops, and being out of reach of the reaper, have ripened and multiplied. Here they spring and wither in retirement, no one caring whether they are dicotyledons or cryptogams. The marigold, the sunflower, and white clover are mixed up with such native flowers as the wild violet, the shepherd's purse, or the blue-flowered "diggers' delight." This latter has come, perhaps, with the seeds from some miner's holding amongst the ironbarks in the gold country, and was once supposed to grow only on auriferous soils. While no one would think of digging for gold in this field, the presence of the flower is, perhaps, as reliable an indication of a golconda underneath as the reports and information on the strength of which many mining companies are floated.

If there had been less gold in Australia, the love of the land would be more the traditional affection of the British yeoman for the soil than what it is. Gold, the greatest of all magicians, is wonderful in its workings, but of purpose most infirm. To-day the prospector's wand is waved above a lonely gully; to-morrow it is peopled; in a week deserted. And this spirit of change has become almost a national trait; not so much in those old settled districts, where homesteads carry the traditions of more than one generation, as in the newly-settled areas, where men sell and pawn their homes. Although Crabbe —a village poet—saw in the occupation of Suffolk reapers only the emblem of the inheritance of toil resting upon the children of Adam, and thanked God that he had been saved from so cruel a destiny, few rustic poets have been so ungrateful to their birthplace.

With all the vicissitudes of bad and indifferent harvests, still the farmer's is a happy life. And though perhaps the struggle of the early days when he first took the land under his control warped his

temper, it only ripened his philosophy. In the
heated competition of the city the lucky ones hardly
even glance back at the friends they have left struggling in gloom. But in the country he who has
fought fairly and honestly, and through no fault of
his own has fallen, will find hands—hard, perhaps,
but steadfast—held out to help him on his way.
Though the fields have little poetry in the days of
toil, in the after time they are a most enchanting
memory. Looking back, with the failures forgotten
in the years of plenty, the troubles smoothed by distance, they say, not with Tennyson's cynic, but
lovingly—

> "'Tis the place, and all about it,
> As of old, the curlews call."

Almost without knowing it they love the land. Even
the very toil that lasted out the daylight had within it
some song of promise. There was music in the gritty
rumble of the plough through the brown mould, the
rustle of the scythe in the rye, or the throb of the
primitive flail upon the canvas of the barn-floor.
Few melodies are sweeter to the farmer's ear than
the thresher-drums at the different stackyards
amongst the hills, their humming chorus softened to
a mere murmur in the distance. In the slack
season of the year there was the garnering of wild
honey. Perhaps for years the bees had been seen
passing in and out through an orifice of the tree. At
different seasons they had different lines of flight.
Sometimes it was to the garden, where forget-me-
nots were in bloom; sometimes far away to the
grass-tree patches on the hill, while the long spikes

had yet broken on one side into a white, frost-like bloom, and on the other the young blossoms resembled white beads set in a background of greenish lawn. Some day, when there is little else to do, it is proposed to cut it down. The old tree cracks once or twice ere the thunder of its fall rolls away through the bush, and the air is thick with the pleasant perfume of the sap—the very heart's blood of this wrecked gum trickling down over its sadly-shattered limbs. The dogs, dashing in amongst the boughs, are stung by the bees, and scamper away. Standing far off, they are moved alternately by opposing impulses—one, the consciousness that there are opossums in the fallen tree; the other, a tingling conviction that angry bees are plenty also. The hollow branches which were the storehouse of the bees are cut open, and the honeycomb carved out in cubes and piled in milk-dishes. Some of it is white and clear—the harvest of the past season; the rest has, perhaps, been stored for years, is browned with age, but full of a rich honey that, like wine, has mellowed with time.

The opinions of the farmer are less liable to change than the city man's opinions. Yet the farmer is in most things practical and considerate. Though he knows that a preacher of old spoke his parables upon the hill-sides of Palestine, he would not insult the lay reader from the city by asking him to preach under the gums. Even before a congregation gathered a church was built. Then the farmer felt that he had done his spiritual duty, and, following the example set by Eutychus, slept placidly through the sermons. The farm hands look at life from quite another point of view. The vagabond life led by most of them rubs all traces of conservatism away. There is too much of smug respectability about most modern religions to suit the farm hand's Bohemian tastes. He reads and thinks generally in one direction. The theory of evolution is rather popular with him, because, as he looks at it, blue blood means only a more direct descent from the blue-nosed ape, and he has no liking for titles or pedigree. Science and religion are with him opposite terms; either one or the other is a fallacy. The orthodox preacher who puts a dash of science into his Sunday sermon is merely seeking credit for breadth of mind under false pretences. But of all sentiments that have existence on the farm the national or patriotic feeling is deepest rooted amongst the men of to-day. The farmer is proud of the kinship between Great and "Greater" Britain, and "the contraction of England" school, who talk of an Australian empire of peace, with separation as the logical sequence, find no encouragement from him. The humanitarian's dream may be a good one, but it must be realised without surrendering those things which his fathers honoured, and for which they fought. Some of them must forget their own names ere they forget that those of their race died for the traditions they are asked to cast away. What matter though the rulers of the old land are out of harmony with them to-day? To-morrow those rulers are gone, but Britain is Britain still. Some of the more prosperous, who have ample leisure to think about these things, go farther still, and reflect upon the price they must pay for a mythical independence. There are ties of sympathy deep and dear. Will they sacrifice the right to bend as reverent children over the grave of Shakespeare, the right to feel proud and fond of the memory of the man who played upon the nation's very heart-strings an accompaniment to his "Christmas Carol?" Nature's better part is often hidden in remote corners, her gems in almost inaccessible places. The most beautiful heaths in Australia grow on steepest ridges of the Grampians. So about farms healthy sentiments cling, I think.

There is an illustration in miniature of the birth, progress, and destruction of a world every summer in the wheat-fields. Planting, growth, and the reaping are a cycle in natural life. Where either plants or animals run wild and multiply they die at length of their own exuberance. When some particular weed becomes a pest the farmer finds it best to let it thrive and take possession of the field, for finally will come a spring when it disappears entirely from the face of the earth. It may be possible that, as with in-breeding, the race deteriorates. If the weed has taken entire possession of the field, the bees and honey-seeking insects move only from one flower to another of its kindred, and so cross-fertilisation with the dust of other flowerets is denied. The republic of the wheat-field is rarely over-crowded, however. The animals, birds, and insects gather and accumulate, and their enemies follow. But before the struggle has developed man steps in with his reaper, and in a few days the community is destroyed or dispersed.

The hares are about the first to establish themselves. Perhaps on some windy night towards the close of winter they come down from the plains to feed, and not caring to face the bitter cold of the plateau again, lodge in the young wheat as well as feeding there. Hares and rabbits attack the young crop in somewhat different fashion. If it borders a river bank the rabbits nibble it down on that side, leaving their traces in a hundred little burrows scratched in the soft mould. But the hares commence on the side next the pastures or the open plain, and cut the blades down evenly. They are less prone to play and run about at meals than the rabbits. They feed round about them, and having taken off everything within reach, move with one long lope forward to higher pasture. Hares seem to have a strong sense of safety in what scientists call "uniformity of environment."

Where fields of red soil have been ploughed hares take up their residence quickly, and are slow to leave their forms when alarmed. It is an interesting study to come within a few feet of a hare thus squatted. It is so flat in the form that nothing breaks the evenness of the surface, but just before the field is harrowed a practised eye may sometimes note a slight unevenness in one of the lines formed by the nicely-packed sods, which induces a second look and a discovery. The chestnut eyes, so closely in harmony with the yellow tip of the fur, stand out on either side, so that front or rear they see you as plainly as you see them!

The pliant ears are flat along the neck, and to accommodate the higher hind-quarters the form is hollowed out a little more at that end. Indecision is the characteristic failing of the hare. It is rarely satisfied with its first form, but after settling down cosily for a minute moves away to another part of the field. So when in motion the ears are back for an instant, and a few strides are taken in real earnest. Then the black-tipped signals, which seem to say "stop her," go up, the hare eases her pace, has, perhaps, one curious look at the enemy while standing on her hind legs, then away again at full speed and out of sight before another halt is made. Living in the young crop, and finding it grow up and shelter her, the hare breeds there; and, just as the season is late or early, the young, becoming venturesome and independent, leave the field when the reaping begins, or lying close as very young hares do, are cut to pieces with the knives of the machine.

If the field border a river with scrubby banks most of the live things which people it in its later growth come in gradually from that side. When the young blades begin to thicken and creep about the ground, as young wheat in contrast to oats always does, there is a perpetual spring dampness underneath. Grubs and slugs are generated, and big green frogs from the river reeds follow in natural sequence, as well as landrails, which feed variously on water insects or the soft-fleshed caterpillars clambering on the young wheat. The wheat springs higher, and the lower blades dying away leave space between the stalks. Then the field mice begin to congregate, and wherever mice and frogs are gathered together there will the snakes come also. As the summer advances the grass is burned away out on the pastures, and the quail of different kinds which live on the open plains are attracted to the crops, and remain there for the nesting season. What with the quail and young hares and water-hens, a pair of the larger hawks soon discover that the field is a splendid hunting-ground, and as they like to camp near their game they at once start to carry things to the most advanced patch—generally a tuft of wild oats—and to weave a bulky and not very comfortable nest. If there are large trees in the field, with a hollow bough broken off, the chances are largely in favour of finding a couple of buff kestrels established there, with their round, rich-brown eggs almost lost in the soft decayed wood of exactly the same tone. The birds hover about the margin of the field, or beat along the furrows, and in the open are a worse enemy to the field-mouse than even the snake. When the field is reaped and the corn stored the mice are taken away with the sheaves, and the kestrels follow them to the stack-yards, and live there through the winter, nipping up every mouse that ventures into the open, and making life eventful and uncertain for the sparrows also.

The snake seems to have a love for some geographical feature about his home. If there is a quarry, a stone-covered ridge, a gully, or a bank about the margin of a field that has been constantly cultivated for hay or grain, a snake will find a home there. When the paddock is bordered on one side by a dusty lane you can approximately fix the residence by so often noticing the curving trail in the same locality where the snake has crawled across into the wheat and back home again. They hunt largely about the borders of the field, and this is why the mice are always more plentiful towards the centre. The snake is unquestionably "mousey's" worst enemy. The kestrel is quicker in seizing a mouse when abroad in the day, but they are not very fond of daylight. The snake follows the mouse home, thrusts his small sinewy head and neck into the burrow, and drags out the owners. The frogs and toads when they get away from the water are a very easy prey also. When there is but one cultivated field by the river the snakes travel to it along the banks, and as they have very much the same architectural tastes, it is not unusual to find more than one in the same home. Once at the corner of a wheat-field in harvest time, I saw no less than five dug from beneath the same dry stump, and two of these were tiger and the other three brown snakes.

The bird above all others closely associated with the wheat-field is the quail, and of the three kinds once plentiful within a few miles of Melbourne two build largely in the wheat, while the other, true to its earlier traditions, clings to the grass lands, even if bare, although sometimes found in stubble. As the reaper comes down the field, felling another swathe in the square of standing corn, you see the quail leave the wheat and run like Guinea chicks down the line of the log fence. The binders following in train find the homely nests filled with a handful of eggs, sometimes slightly under, sometimes over a dozen. These belong to the partridge quail, whose eggs are always of one colour—a bluish white lightly speckled with yellow, and sharply pointed at one end. The young are rarely seen, for, like most of the plain wanderers, they have a wonderful knack of hiding where there is no apparent shelter. The colour of the partridge quail is brown lined with dull white, and its yellow legs, as well as the freckled eggs, are very much in harmony with the stubble. The pectoral quail lays occasionally, but not so often as the partridge quail, in the wheat-fields. Its eggs vary most remarkably in colour, so much so that anyone taking them as a guide to species would be led to believe that at least a dozen varieties of quail were native to the same locality. From a cream ground heavily blotched with chocolate-brown the colours range through a score of tints, principally greys and browns, down to that universal dusting of colours which in tweeds is popularly known as "pepper and salt." The third species of quail, more often found on the plains than in the corn-fields, lays four large eggs—the underground buff, thickly dotted with ashen-brown. The nest is a mere hole scratched in the earth, and however often one may have startled the quail from the grass, and been startled by them in return, as they sprang with a burr of wings from almost under foot, the birds were yet rarely found near home. To-day, as you walk along, your eye rests by chance upon the single egg lying in the little hollow. To-morrow, as you pass, there is a second egg, and finally the customary four, but come as often and as carefully as you may, you never find the mother bird there. Often the eggs are warm, showing that she has just left them, but the quail seems to have the gift of lying flat as a flounder upon sand, and must be constantly on the alert. She probably runs from the nest as soon as a footfall is heard in the distance or an object is seen upon the plain. Sometimes the nest of eggs is the only evidence that there are quail in the locality. The rarest and most interesting of the grass quail is the plain wanderer, more timid of aspect, longer in neck and leg, while generally more awkward also than the other species. Its egg differs from the others in being less sharply pointed, and the bird takes flight so reluctantly that an adult quail is often mistaken for a fledgling. Unfortunately the two most prolific breeders among them build so often in the hay-fields, nests deserted and eggs addled year after year must thin them down greatly. They fared best in the wheat, for where the stubble was long they were not disturbed. When fields are planted for hay the land is rolled in the spring until not a clod remains, so that at harvest time the mower may shave close down to the ground where the cornstalk is thick and weighs best. What the mower spares the hay-rake is certain to destroy. The landrails suffer considerably also by building in the fields. Their large white eggs with flesh-coloured spots are much more easily seen when exposed to the daylight, and are taken by the first predatory bird flying over freshly-cut stubble on the look-out for these tasty morsels. The landrail, with its black-barred breast, beautifully marked wings and back, tilted tail, and bright red eyes, is the smallest and certainly the handsomest of Australian coots. The name "water-hen" is given in the country to all the coots. Out in the West the porphyry coot is the water-hen, but down about the Gipps Land lakes the water-hen is the little black coot, while amongst the hayfields round about the Werribee and Keilor plains the landrail is rarely called by any other name.

Even in Australia we have had three eras in the harvesting of wheat—first, the days of the sickle, the cradle, and the flail; then the reaping-machine, the "sheaver" with his rake, and the train of binders, expert in the mysteries of the "gooseneck" and lock-band, following the reaper round about the field. Last of all came the stripper and the reaper and binder, overturning all the old ideas, and taking all the poetry out of field labour and harvest time. At this period of the year in the country, when even the brightest noonday is followed by a chill, grey evening, and the wreckage of the dahlias and chrysanthemums yet clinging to their stalks along the garden-paths are the only mementoes of past glory, one looks pleasurably forward to spring and summer, to seed-time and harvest. Anyone who has once seen Australian plains in the early summer—the flowers lying in belts of orange, blue, and pink, the patches of sorrel in the fields, and the tufts of bronzed summer grass waving silkily as a child's tresses in the breeze, must long to see them again. The early almond trees just breaking into blossom show that Nature's panorama is about to open out again, and a new community to be gathered together, fostered, encouraged, and then, in the very prime of its existence, cut down and gathered with the wheat.

Kingfishers and Pigeons.

The best known and most thoroughly Australian of all the race of kingfishers is our sober brown friend the laughing jackass. The name of kingfisher is hardly applicable, for he is no fisher bird, and his plumage, naturally dull in colour—and about the city blackened, like the sparrow's garb, with floating smoke and dust—is much too homely to be in any sense the raiment of royalty. The laughing jackass is a stay-at-home bird. All through the winter I watched a pair of them in a small clump of willows in the Fitzroy Gardens until the trees were draped by that magical milliner, Spring, with streamers of tender green. When the birds first took possession it was a pleasant summer arbour, but although the curtains browned and fell slowly away in the autumn, leaving nothing but skeleton boughs, they were too loyal to forsake it. Here they stayed in all weathers, when the rain streamed down upon their shelterless home and the lightning lit up the bareness of the avenues of elms and planes. Sometimes they were fossicking on the ground amongst the dead leaves, their dull coat lost in the harmony of brown. The damp leaves were a harbour for worms and snails, so that food was always plentiful. In the nesting season they will leave temporarily for some hollow gum tree, where, like the sparrow hawk, they lay their eggs in a nest of soft, decaying wood, but this triangular enclosure of willow trees is yet their empire. Watch one of the birds perched upon his bough in broad daylight. He seems to be either asleep or lost in deepest meditation. A worm crawling beneath the leaves shows just the merest speck of its glistening skin through a rent in the rustic carpet. Suddenly as a falling meteor the laughing jackass drops upon it, and, flipping aside a bronzed leaf, throws up his bill with the wriggling prize. Should a frog, alarmed by the intrusion of a thrush or minah, leap from its shrubbery retreat into the daylight, the dart of the sedate sentinel on the willow is equally quick and quiet. His tail tilts victoriously as he pinions the unlucky frog, and in this sprightliness, compared with the seeming stupor of a few moments ago, a sort of malicious glee finds expression. His low chuckle seems to say, "You thought I was asleep. But not much." One night at the entrance to the gardens something flapped away on soft wings overhead, and I thought I had caught my friends of the willow clump on a night raid. Passing the favourite haunt, I was able, by bending low to the walk, to "moon" the pair sitting reflectively as ever on the familiar perch. The night hunter was no doubt an owl. The call of the laughing jackass is less the laugh that so many have been pleased to fancy it than a varying owl-hoot. There is nothing fiendish, nothing maniacal in the rolling note rising and falling and echoing between the trees at sunrise and sunset. To the bushman, if not to city people, it is even musical. None of Marcus Clarke's fine fancies were more unreal than when he spoke of that bush chorus as "horrible peals of semi-human laughter."

The nearest relative of our southern laughing jackass is the bird known to naturalists as Leach's kingfisher—a sort of improved jackass. In shape and size the two are much alike, but residence under the brighter sunshine of North Australia has much improved the plumage of Leach's kingfisher. The main ground of his costume is lighter than that of the southern jackass, and while in the latter the neck feathers are just tipped with a darker shade of brown, the other has a distinctive hackle streaked with black, and not unlike that of a light Brahma fowl. Through Queensland and in New Guinea his markings are very beautiful. Above the tail and on the shoulders of the wings are patches of the most delicate of lustrous light blues. The silky tint is more nearly akin to the lighter blue of the precious opal—found with the bird itself along the banks of the Barcoo—than any other colour in nature. The association of this blue with a rich chestnut brown, while in effect not comparing with the brilliancy of the parrots or some of the smaller tropical kingfishers, makes yet a tasteful and striking costume. The jackass found along the Murray is a connecting link between the two kinds, just as the halcyon of the same river fills the gap between the azure kingfisher and the halcyon of the "back blocks." A few pale blue feathers appear in the wings of the Murray jackass, and the brown in his coat is more brilliant than with the dweller by the Yarra.

The true Australian kingfisher—for although several of them are strictly water birds, this one is the more universal—is the azure kingfisher. Who has not admired that streak of azure—brief as a lightning flash—thrown for an instant across a yellow winter pool, or the shades of deep orange and fawn red reflected by the clear waters in midsummer as the bird darts across? Often on the shaded side of river groves, angling for stream trout, I have been startled by his sudden dart past, or have espied him in some quiet corner, sitting patiently on his perch above the stream, just as our garden jackass sits on his willow bough? Suddenly he, too, darts down beneath the water, and scattering the liquid diamonds from his wings as he rises, is back with a silver atom of a fish flapping feebly in his long black bill. If after a dive the birds fly off with the fish, you may conclude that there is a young family not far away, but the nest, generally in a bank, is not easily found. Though as a boy I knew several localities where it was almost a matter of certainty that a pair of kingfishers had their home, I only chanced to find one. Climbing one day along the limb of a fallen tree that overhung the stream and inclined gradually upward from it, I bent over to peer into a hollow at the extreme point, where it had been broken sharply off. It was tenanted, for a grey mopoke within a few inches of my face suddenly opened his wings, eyes, and mouth all together. This apparition with the gaping yellow mouth was so startling that without delay I dropped into the water and swam ashore. Climbing the bank, a blue kingfisher darted out, and the interesting nest was revealed. The mouth of the hole was partly hidden by long strands of grass that from a tuft above hung down and screened it. At the end of the tunnel were three delicate white eggs, the yolk shining through the shell with a faint reddish tinge, as it does in the fragile eggs of the yellow-tailed wren.

However they may differ in plumage, the kingfishers are alike in one respect. They all build in a hole, either of tree or bank, and lay white eggs, rounder than those of any other Australian bird. The parrots, which also build in the cleft of a tree, have distinctive white eggs, but the shape is a perfect oval. The floorway of the nest of this azure kingfisher was paved deep with broken fish-bones, and the hole must have been an eligible family residence for years. How many fishes had contributed their skeletons to complete this avenue? On top the bones were snow-white. None but a water bird could have kept its walks so clean.

There is a distinction, subtle, perhaps, but always discernible, between the true water kingfisher and those birds of the race that are often found far from the streams and living almost without water. Take the azure kingfisher as a type of the true water-gnome—the bird holding the place amongst his kind that the water-lily does amongst flowers. His bill is longer, darker, less conspicuous, and more elegant than that of the bush kingfisher. Between head, neck, and body there is no distinctive break. They glide imperceptibly into each other, snake fashion, as in the bittern and some of the cormorants. The snake-bird of the Murray, with its long sinuous neck, illustrates to perfection this peculiarity.

With all deadly reptiles, however, the eye is long and narrow, while the eye of the kingfisher is round, jet black, and bright, as befitting a bird which depends for its living on the power of out-seeing and out-swimming the fish in their own element. As the azure kingfisher drops to the stream his bill, head, neck, and body form a straight line, and he shoots through the water like an arrow, leaving scarcely a ripple on the surface. It is the flap of his wings as he rises again which attracts our notice. He has no tail worth mentioning, while his cousin of the bush has a few terminal feathers which droop despondently or are gaily tilted just as the bird may be sleepy or very wide awake.

The kingfisher at home is always an object of

interest, but how much more so in the far interior, where his whistling note is often the only bird-sound that breaks the impressive silence, his bright plumage the one speck of colour in the waste of withered nature.

The bush halcyon is to be found amongst the Central Australian myalls all through the summer. The streams and tanks may disappear, the flocks perish, the wild birds and beasts seek haunts where Nature is less harsh, but when, after a rescuing shower of rain, the boundary-rider revisits the "White Plain Paddock," he finds that where all else has yielded to the sun and the drought the halcyon survives. Like the shepherd himself, it endures this lonely life. There is a link of sympathy between the halcyon and the shepherd. The man dreaming, as that other shepherd Endymion dreamed upon the hills of Latmos, may see some homely face, some fair divinity of years ago, mirrored in the splendour of the moon, but his companion by day is the little halcyon. It is as conspicuous here as a sundew in the grass at noonday.

The kingfisher of the plains is like the sacred halcyon in some respects, but its colour is much richer. The red on its back is a deep chestnut, with the glow of polished jasper, but with a warmth and softness never seen in any colour of stone.

The sacred halcyon is more often met with. I have found him all through the north of Victoria, and, although not a water bird, he has a special liking for the tortuous banks of the Broken River and other streams on that side of the Dividing Range. Again, where the Parramatta narrows from an arm of the bay into a river the halcyon may be seen darting about the margin of the mangroves, following down the receding tide, and feeding upon the marine life left stranded on the sandpits. The distinctive trait in South Asian varieties is a sprinkling of many colours, while broader patches of a single hue characterise most of the tropical kingfishers of Australia. The kingfishers of North Australia and New Guinea are a wonderful collection, and they have imposing scientific titles. The popular names of the South are much too homely to fit these splendid birds. One of the halcyons has a snow-white breast, with a rich mantle of indigo thrown over its back, head, and wings, and the merest suggestion of gold trimming where the blue and white unite. Another has a pure white head, neck, and breast, but elsewhere a radiant green —merely two colours, yet how complete the costume! It is not, however, in the glory of colour alone that Nature has been indulgent to the northern kingfishers. One of them has a coronet like a peacock. Another, impelled to ambition, perhaps, by jealousy of the birds of paradise, has thrown out from its tail two long splint feathers. Should North Queensland ever get separation, the state museum might contain such a mass of sparkling colour as no other colony ever gathered from the life of its own forests. But the Queensland bush is, after all, its best museum. Many a man who has fossicked in Nature's byeways has at one time or another believed that the chief pleasure in life was to shoot the rarest birds, stuff them with cotton-wool, pierce them through with pliant wires, and twist the fading skin into a crude imitation of life. Suffering from such enthusiasm I once fired at a kingfisher as he lit upon a broad stump, where the waters foamed over a bank of gravel. When the shot struck him he just spread his wings in a dying spasm. As I left him lying for a time so that the plumage might not be dabbled with the blood, than which it was not less bright, a first sense of the cruelty of the whole thing dawned upon me. How I should miss the familiar flash across the waters in the evening as I strolled along on the look-out for a hare or a family of wood-duck on their favourite gravel-banks. It is, indeed, better to leave the gems of the forest in their natural setting.

Amongst the kingfishers that range westward from Cape York two are absolutely the smallest of their race. One of them has clearly-defined waves of light along the neck, like the watery sparkle in the plumage of the English kingfisher. The little kingfisher, not larger than a red-backed finch, and more shapely, is another of the brilliants of the bush. Its breast is flaked with white, and the back a pure indigo, while the sharp black bill and tiny white legs are the tribal marks of the true water kingfisher.

The wild pigeons of Australasia have a range in variety as wide almost as the fancy birds that breeders have brought from all corners of the earth. One can't help thinking that they are misnamed. What affinity can exist between the great crowned pigeon of New Guinea, in its garb of slate-grey, white, and bronze, topped with an airy, fragile crest, in size almost a bustard, and the puny wren-pigeon of the torrid "northern territory"?

Yet greatly as the pigeons vary in general appearance, they are alike in that peculiar smoothness of plumage, that roundness of outline, and that softness and simplicity of expression which have made the dove a type of affection, fidelity, and innocence. Some of our wild pigeons may tend towards the toucan tribe, others show some of the characteristics of wading birds, and still another section verge very closely indeed upon the quail family; yet in character there is everywhere a communion amongst pigeons not easily mistaken. The most beautiful of all the Oceanian pigeons are the doves of Fiji. Those islands are themselves amongst the beauty-spots of the Pacific, and their doves are the gems of the luxuriant woods, for, though less elegant in shape than most of the mainland birds, their colour is very striking.

The orange dove has its name from the depth of that particular colour in its plumage. Indeed, the tone is too rich almost for the title to be strictly appropriate. The redness is rather that of a marigold, and one of these birds amongst banana and palm trees is as conspicuous as these old-fashioned flowers against a bank of ivy. The plumage of the golden dove has a main-ground of olive-green, striated with gold, and its luminous pale pink eyes are expressive of great tenderness. The nutmeg dove of the same island—one of the fruit pigeons—is worthy of rank with those two other stars of the pigeon world.

The fruit pigeons of the north-eastern corner of the continent are among its most interesting zoological studies. They harmonise with the rich inflorescence of these summer forests, and yet in their plumage are few primary hues. The pigeon-green is not the emerald tint of spring meadows, nor the darker gum-tree green of southern forests, but rather resembles such indistinct shades as myrtle, sage, and olive. The fruit pigeons' beauty of form and colour is impressive, and their eyes of ruby and yellow are indicative of a depth of colour, rather than of the sparkle of the jewel. Standing at the head of its order is the magnificent fruit pigeon, its head pale grey, breast shaded in rich plum-colour, and sparkling on its back iridescent tinges of gold and bronze-green.

As it clings with long, pliant claws amongst the pendent sprays that in the upper strata of Queensland forests sway about like the feathery plumes of the white birch, it is a rare gem among brilliant birds.

From this bird to the Torres Straits pigeon (nearly white, and with just the suggestion of an Arctic hardness of aspect), there are many notable varieties. In the home of the Torres Straits pigeon—a nest of a few cross sticks and a single white egg—there is a reminder of desolation. Nearly all the fruit pigeons have a characteristic head-piece—in the majority rose-pink, though it varies through different shades of crimson, pale-pink, and grey, ending finally in pure white. The gradations are so faint, that though two birds chosen at random are unlike, intervening species will be found bringing them gradually into close affinity. All the Queensland woods are the home of some one or other of the fruit pigeons, and there alone they are seen to full advantage. Whether feeding on the Moreton Bay figs or the score of other native berries, brighter in colour than themselves; whether on the fronds of some lofty forest fern, flapping against the white-flaked leaf of a caladium, clinging to the tendrils of a flowering Tecoma or a Wonga vine, or wafting the perfume of musk plants through the still forest with the beat of its wings, the raiment of the fruit pigeon is ever in harmony with the radiance of the scene.

None of the Victorian wild pigeons at all compare in beauty of plumage with the sub-tropical birds of the same order, yet they have an interest of their own, and a very wide range of distribution. Diverting little wood nymphs are the peaceful dove and the ground dove, both of so mild and confiding a nature that they seem in time to enjoy man's companionship, and are both so easily tamed that they become rare toy pigeons.

The crested pigeon likes the far north-western corner of the colony, where the Murray curves slowly northward for its final run to the sea. There they cluster in sociable hundreds, and become an easy mark for the stockmen and river voyageurs.

But the two varieties of wood pigeon best known all over Victoria are the Wonga and the bronzewing. Both are darker down here than in northern areas, and nowhere are they found to such perfection as in North Gipps Land, where the sportsman has not yet thinned them down.

Among the great stringy-bark forests to the north of the Lakes' Entrance they were common until quite lately. They have been followed as far as the rich maize flats on the banks of the Snowy River, but beyond that, and away up to the Manaro, Wonga and bronzewing enjoy an unbroken rustic peace.

It is a misfortune to both that they are estimable table birds, and so neither their love of solitude nor the density of their native thickets save them from destruction. Once idling away a week in farther Gipps Land I had an opportunity of noting something of the habits of these birds of the shade. On the long slope of a sterile range they kept company during the day with the lyre-bird and the wallaby, but at evening the bronzewings came flitting down to the one creek that drained this long stretch of woodland. This stream, idyllic in its summer clearness and beauty, is a wild torrent in winter-time. Its bed is filled with dead trees, their trunks up stream and the branches downward, just as they swung in the water when the winter flood water receded and left them stranded. Along the bank of the little stream the sunshine comes in rare patches, and even the heavy rains only filter through the thatch of forest leaves. Here and there where flakes of yellow light fall they are broken by the shadow outline of fern leaves upon the bank. The sandy sides of the little neck seem more yellow by contrast with the dulness, and out where the forest opens a touch of whiteness is given by the flake blossoms of the native elder. A bird by the stream has a sweet rolling note, like the song of a thrush. On one of the pools a number of raft water-lily leaves, with the edges turned upward, rock like shallops in the wind that strikes this corner of the pond. A few wild convolvuli, bearing pale pink flowers, striped with white, cling about the brambles, and seem to have been designed as dew-cups for the birds. Down to this stream, after sunset, the bronzewings come to drink, the click of their wings in the stillness of the deep valley announcing their approach. They pitch about the outer pools, and sit for an instant on logs and pinnacles of rock as though cast in bronze. Excepting, perhaps, the crested pigeon of the interior, the bronzewing is the fastest flier—the carrier amongst Australian pigeons. A fast-flying pigeon may always be identified by the rig of its wings—the shoulder thick and clean, and standing out loosely from the body. It must have some of the length of pinion that marks the wonderful passenger pigeon of North America, feathers sharp and well set together, without any of the ragged softness of the laughing jackass or the mopoke, whose flight, as becomes night-birds, is slow and soft. The quill of the feather is thin, clean, and with the hardness almost of ivory.

The pigeons of Australia are of two distinct types—the exotics or fruit pigeons, and the brushwood birds, the latter, in some respect or other, generally allied to our bronzewing. The brush bronzewing is hardly a variation of our pretty bird, its appearance, nest, and habits being almost the same. The South Australian bronzewing is plumper, but hardly so bright, and away amongst the beautiful scarlet-flowered gums of the western coast of the continent they build their nests in the heart of radiating spines of the grass tree. Up north the overlanders find them with well-defined tints of shining yellow and green in the wing feathers. In the thickets where the bronzewing lays its two snow-white eggs the bird may congregate in flocks, but they are not easily picked out of the general darkness of the scrub. When food is most plentiful the bronzewing becomes the plumpest and, in appearance, the most tempting of Australian table birds. Excepting a young curlew, and, perhaps, snipe and quail cooked by an expert, there is no bird of finer flavour than the bronzewing. An old swan is literally a rank failure at table, but the young bird, just before the pin feathers sprout, is "a dainty dish to set before a king." Another splendid table pigeon is the Tasmanian brush bronzewing, found in the sterile north-eastern corner of the beautiful island. At dark its peculiar call is heard echoing amongst the blue gums, giving an added melancholy to these lonely forests. The hum of the pigeon's wings when it takes flight is as sudden and almost as startling as the rise of a quail underfoot. Another of the table pigeons is the partridge bronzewing of the interior, the cooked flesh of the bird being very white and of splendid flavour. It combines the peculiarities of two distinct types of birds. It is a ground bird like the quail, feeding amongst the white summer grass and dead wood. It builds on the ground, lays eggs which are not altogether a pure white, a most distinct departure from pigeon traditions all the world over, and when its young are hatched they immediately leave the nest and run about like chickens. It is hard to decide at first glance whether the partridge bronzewing is really a partridge or a pigeon, and hence the vague combination of title. The white willow grouse of Norway has a head not at all unlike some of the short-faced fancy pigeons.

Some of the little round doves of North Australia are as distinctive in their markings as the bleeding heart pigeon. These pigeon midgets are marvellously tame and easy to approach. One has an erect pliant tail, which, when a glimpse of it is caught bobbing amongst the bushes, will remind one of the tiny blue wren, or the black and white wagtail sporting by southern streamlets. Unique amongst Australian pigeons are the flocks of harlequins that feed on the plains of Central Australia, in the plumage of which patches of bronze, red, slate-grey, and black are peculiarly commingled, and the pheasant pigeon, which has some of the striking characteristics of the Argus pheasant. Pigeons may vary in colour and form, but their eggs are always pure white. The eggs of the different variety of hawks are generally a rich brown tint—no two kinds are alike ; in the eggs of the quail the same variety in colour is noticeable, but with pigeons it is always a pure white ; and as the birds nearly all build exposed nests, one particular theory as to the origin of the colouring of birds' eggs is somewhat shaken.

The Home of the Blackfish.

MANY bits of winding creek, haunts of the wood-duck and kingfisher, might be explored under this title. The blackfish has his home amongst some of the most idyllic forest scenes of Southern Victoria. What a rush of bush harmony the very name invokes! The brushing of leaves, the echoing call of mountain birds, the mimic thunder of winter streams, or the pleasant tinkle of summer fountains. I have found this truly Australian fish in Otway valleys, where one realised for the first time the meaning of solitude, and was oppressed with the consciousness of being utterly alone. In pools where a hundred ferns stooped to drink, and across which a hundred ghostly white-trunked gums threw shadowy outlines darker and denser than themselves. In Gipps Land, when the waters were ruddy brown with the healthy tinge of rich forest loams, or black with the filtered essence of mountain fires. In quiet pools by Pastoria, where every scented breath of summer brought down a hail of wattle blossoms falling without sound or even the suggestion of a ripple upon the smooth water. Yet farther north, where the rivers that crept lazily over the plains were marked in their course only by stiff avenues of red gums, and the banks were without the rich robes that partial Nature gives bounteously to the rivers of the south. I write to-day of a creek half rural, where English cresses mingle with the water celery, and alien willows jut up between the lightwoods. Here in the evening the chuckle of the laughing jackass is echoed by the bleat of sheep winding up in long lines over the hills, and at night the challenge of a farm watch-dog is answered by the melancholy stone plover, whose pipes sound shrilly up the river valley. This is the home not of the blackfish alone, but of those children of the bush—feathered, furred, and foliaged—which, all holding power to charm human senses, are not alarmed at the human presence. The day we spend upon the stream may be that on which the fish are coy and hard to please—all fish have such days—but it rarely happens that fish and flower, animals and birds all seek retirement at the same instant. Indeed, the contrary is always the case, for the eye of the angler, no longer chained by duty on the pool, takes in the wider range of opposite bank, notes every glint of a wing, every whisk of a tail, every leaf that falls rocking to earth, and every scrap of flower colour in the long stretch of brown and green woodland.

The orchard grounds slope towards the river until the limit is reached in a long row of spreading apple trees. There the bank drops more abruptly to the water, and between the apple trees and the stream is willow land. Long ago the willow slips were planted here, and—like the apple trees—being neither cut nor curbed, they grew as Nature wished. The gnarled arms of the apple trees, like those of a forest oak, are flung abroad and aloft, and the branches cross each other and intertwine so as to make quaint patterns against the sky. Even from the hill-top the contrast between the broad bright willow ribbon and the dark-green line of apples is sharp and distinct. There is no blending of the two colours. But underneath the willows there is variety enough. On the lower side, next the stream, the branches by some magnetic attraction are all drawn down to the water, where they send out a hundred bright pink suckers, just as the branches of the banyan on drooping to earth send out fresh roots. Drawing sustenance thus through both root and branch, it is little wonder that the willows are tall and thick and green. The sunlight comes through in fragments only, working a pattern on the turf or the bronze carpet of fallen leaves, like the stains and blotches in a slab of marble, but always changing. Now, in the autumn, the apple boughs are thinning, and the willow leaves are floating away in yellow fleets upon the bosom of—

> "—— the brook that loves
> To purl o'er matted cress and ribbèd sand,
> Or dimple in the dark of rushy coves,
>   Drawing into its narrow earthen urn,
>   In every elbow and turn,
> The filtered tribute of the rough woodland."

The long green arbour is always full of life. Even the hares come down from the hot plains to make their forms amongst the cool green grass by the waterside. The banks are everywhere drilled with rabbit-holes. Watching the little ones skipping about, one would never imagine them capable of becoming a national scourge. I have passed many an hour lying on a grassy bank or a moss-grown tree-trunk within a few feet of the burrow, in order to see the little rabbits at play. To-day they were probably out before we came, but the sound of a footfall on the turf, or the rustle of the prairie grass as the sharp-sounding seed sheaths were brushed together, sent them scampering home. Although the young rabbits have an hereditary dread of man as an enemy, they have not yet learned to pick out his figure from amongst the other objects about, so that it is safe to lie quite close to the burrow as long as we remain perfectly still. One rarely sees the first rabbit appear. Perhaps for an instant I had been watching the water-hen wading in the rushes, and when my eyes turned to the burrow again there was a little head just showing, and a shining black eye which seems to see all over and through one.

The long ears and bright dark eyes of the rabbit are very conspicuous. They repeatedly betray him when his brown body would otherwise have escaped notice amongst the rocks or dry grass. If hiding, both rabbits and hares have their ears lying flat, but their eyes must remain open, and the large brown orbs of the hare are not nearly so noticeable as the black eyes of a rabbit. Indeed, the eye, although so small in proportion to the rest of the body, is remarkably conspicuous. Many reptiles which lie in wait for insects on the sloping trunk of trees trust to escape observation because their rough skin is exactly the same colour as the bark. They lie with their eyes almost closed. There is just a faint bright line perceptible between the lids, but the eyes open and flame up viciously when danger threatens. Were they always wide open the hawks would find foraging a much easier occupation. Birds that hunt at night, such as mopokes and nightjars, have the same peculiarity. Were their eyes as wide open in daylight as those of other birds they would less frequently be mistaken for a piece of dead wood left in the bushes by the last fresh in the river. Before this another rabbit has joined the little fellow we left on guard, and the pair hop out to their playground—the well-beaten mound of earth a few feet from the burrow's mouth. Here they are joined by the rest of the family, and the fun commences. Perhaps one of them is so venturesome as to stray quite ten feet away to one side, but the rustle of a falling leaf or the flutter of a wing in the bushes brings him back again. "Birds in their little nests" must, for obvious reasons, agree, but young rabbits, inoffensive as the race generally are, do nothing of the kind. It rarely happens that the assembly breaks up without a fight. Perhaps the bully of the family attacks a weaker brother, and their nervous little fore paws are used with a will for a moment. One of the family is always ambitious to climb the bank above the burrow—a feat attended with many failures. Another perhaps finds a patch of sunlight falling through a break in the willows, and stretched there at full length, with his ears lying back, he looks very sage and comical. If you are anxious to see how young rabbits act on a sudden alarm you have only to raise your hand sharply. For an instant every little fellow falls flat upon the earth, as many puffs of dust are kicked into the air, there is a twinkle of white tails in the portal of the burrow, and you are alone. In less than half an hour you may depend upon again seeing the little bullet head and bright eyes of the pioneer at the entrance.

The song of birds is not the only music heard under the willows. On a bright day in early summer the tree locusts make their presence felt. In the choir there is variety of sound as well as species. On the branches of the taller trees the black locusts with ruby eyes and semi-transparent wings are the baritones. They have a varying call—at first a regular bagpipe, then two distinct notes frequently repeated. The volume of sound is perhaps greater than from any other insect of the same size, and seems to be assisted by a drum in the lower part of the head. The tenors range in colour from bright green to pale buff. The former predominate, but very few specimens are found of exactly the same colour and with similar markings, although hybrids between black, yellow, and green are common.

One great merit in the willows was that they permitted some of us to enjoy a couple of hours' quiet fishing on Sunday afternoons without offending those who choose to take more gloomy views on the question of Sabbath observance. Not that the rustic Puritan is as rigidly respectful of these obligations as he pretends to be. He certainly does no manual labour, but, under pretence of a Sabbath walk, he strolls amongst his cattle, and his crops, and his fruit trees, and plans out the labour of the following week while gaining credit for serious meditation. But if you have no cattle and crops to inspect, and prefer to spend Sunday by the river-side, listening to and admiring Nature, the great heart of the Puritan is wrung with concern for your hereafter. Here in Willow Land you hurt no feelings save those of the fishes, and it is the deed and not the day to which they object.

The blackfish is of stay-at-home habits. You cannot easily tempt him from his favourite river cavern into the broader stretches of water. He must be sought in shaded nooks and retired crannies of the stream, and at those points difficult of access where, it may be presumed, none but the most energetic angler has ever cast a line. In trying for blackfish the fisherman's art is not so much a choice of baits, the delicate manipulation of rod and line, or the hundred and one well-planned deceits that in the more spirited quest of trout and salmon show the relative value of the human intellect as compared with the instinct of the fish. The first essential to success in the pursuit of the sluggish Victorian fish is the power of estimating the arrangement of logs near the river bottom and the curl of the undercurrents from what may be seen upon the surface. This faculty comes only as the result of long experience, so that while the skilled angler drops his bait within a few inches of the reposeful fish the lure of the novice is lost in a Sahara of open water. The haunts of the black-fish are more easily found than described. Perhaps the important topographical feature is the huge black bole of a long-dead gum tree lying parallel with the bank, and one end just cutting into the current and diverting some of its filtered tribute into this little bay. If the trees—not willows or others destitute of winter leaf, but native ever-greens—droop over and shelter this tiny cove so that only a single hand's breadth of break in the foliage remains, so much the better. If there is no such opening you may snap off a bough to-day and drop a line there a week hence when the fish have become accustomed to the changed light. They are especially active during that quiet and too-brief time between sunset and darkness, and so it happens that the angler's skill in finding the home of blackfish determines his success. If a cover is drawn blank once or twice in the evening, that magic time, the dinner-hour of the fish, is wasted. As the darkness falls the float lies motionless for a time, and the angler awaits the coming out of the eels. They are late in leaving their nests in hollow logs, amongst reed-beds, and underneath boulders, but they feed far into the night.

In that stillness between the retirement of the one fish and the coming of the other, night voices, unheeded in the anxiety of sport, force themselves upon the notice. As the angler sits in the darkness behind the shaded fire or lantern he is himself invisible, but the light throws a broad, illuminated band across the water, just reaching to the boughs that dip to the stream from the opposite bank. A platypus paddling gently with his webbed fin-claws drops quietly down with the current into the band of light, and if a hand is moved dives sharply—the splash very different from the ripple left by a stealthy water-rat. The water-rat is something of a traveller. An artificial lake was sunk in the centre of a village near here, and though half a mile distant from the river, the water-rats soon made it their home. The opossum has an odour of his own; an accented flavour of the gum leaves, amongst which he lives and feeds, which also taints the cooked flesh, otherwise white as a chicken and pleasant to the taste. You know that the opossum is abroad long ere he goes leaping up the gums, dry bark shreds rustling behind him, and before his indescribable throaty call breaks in upon the silence. The opossum likes the river trees because there is always a variety of food there. In the bends, where young crops such as maize and wheat are growing, or immediately beneath the orchard, the gums are always scarred with the tracks of many opossums. Amongst fruit trees at night you find them taking a share of apples, pears, peaches, and cherries. His sharp nails enable him to go leaping in his peculiar way—head and fore-quarters well down—up the straightest, smoothest gum tree. The native cat—which is not a cat but a swiftsure—has blunt claws without a sheath, and can only climb a gum tree when it slopes considerably, but is very much at home in the rough-barked sheoak, its favourite night haunt. "Mooning" opossums is a speciality with country boys. The juvenile hunter utilises the moon as a cavalry patrol would his field-glass for every suspected point. And when finally located in the full centre of the moon, the hairs on the back of the opossum seem to stand up separately, so distinctly is every minute object brought out by the luminous background. Against the moon at night I have made out a wicker-work bird's nest which would have escaped notice altogether, from the similarity of the surroundings, in daylight. Sitting thus in the "field" of the full moon you see the opossum in his slow, impressive way, and with those large, bright black eyes, making a placid inspection of the invaders of his home, yet fearing no danger in his lofty perch. The monkey bear cries pitifully when the axe strokes ring out at the base of the trees in which he is perched, and his terror increases while the red chips fly about, but the opossum is less apprehensive of harm.

Happy that opossum whose home tree is so chosen that the enemy cannot bring its branches fairly between his eye and the moon. The moon is really the opossum's worst enemy. It assists not alone in finding but in shooting him. With a fragment of white paper or chalk-mark on the sight— or, better still, with the new "diamond-tipped" sights which science has suggested to our modern gun-makers —one can hit a dark object at night provided it can be separated ever so slightly from the surrounding blackness; but with the silvery moonlight pouring down over opossum and gun-barrel alike a miss is hardly possible. A novice may fail through the mistake of looking down upon the light-flooded barrel rather than along it, and so send the charge humming over the back of his game, just close enough to drive it scampering home. Opossum shooting is sport for boys only. When bushmen shoot or snare them in the frosty nights of midwinter it is for profit rather than amusement. The fur is then thick and tipped with silver—a beautiful contrast to the dull, patchy, brown coat of midsummer—and works up into very handsome rugs.

The angler has time to study all this night life, and much more, before his white float (it should always be white for night fishing), which has long rested motionless, like the sails of a becalmed ship, sinks suddenly, without a tremor, into the water, and he knows the eels have come. Blackfish and eels take the bait differently. The former carries the float along the surface of the water in a succession of runs towards home, and you may get much nearer the heart of his family circle in the next cast by noting the direction. The eel, on finding an imaginary morsel, goes straight for shelter to the river bottom.

One often wishes that he could have spent his youth in a dozen different places, so that he might have become familiar with the wood magic of all. But even when confined for observation to one bit of bush land, we fail to realise how many traditions of Nature are stowed away in the store-house of the mind until, by some chance touch, the key turns and the door swings open. Here in the treasure-box are the old-time incidents scattered before us fresh and unhallowed by dust or decay, forming of themselves a picture in which youth, vitality, and pleasure are limned in tints indelible. Surely some good magician controls what Tennyson calls "the dewy dawn of memory." Seen in the faint light of this dawn our fishing failures are the excursions best remembered. When the blackfish ignored us altogether we gave attention to Nature with better reward. It was annoying, perhaps, to bend down over the logs, and with our eyes accustomed to the water to find dark objects swimming sluggishly in the depths below, and yet resisting every overture in the way of tempting baits. Spring is the angler's time of pleasure. Then the knowledge that in the yellow murky waters of the lingering winter freshets the blackfish and eels are abroad all day tempts him as a constant challenge, even while the paths leading riverward invite him to linger. The meadows on either hand are covered far and wide with a yellow mantle of cape weed. The flowers rattle against our feet as we stroll, and shed over us, as upon the companion scarlet pea, the burnished ranunculus, and the lady's cap orchid, a lustrous dust. Lie down upon this velvet couch, and under all the voices of the fields and hills there is a low monotone, a whispering music, that can be mistaken for nothing else than what it is—the humming chorus of bees. It is fainter, more musical, less melancholy than the murmur of sheoak plumes in the breeze. In the growing summer, the time of flowers, the bees are all abroad, each working out its little destiny. These are Ligurian bees, emigrants like ourselves, and not the long-bodied natives one finds buried waist deep in the floral mausoleum of the yellow box. Bees have always loved the sunflower, and the Cape weed blossom is a miniature sunflower both in appearance and habit. There is the same tiara of yellow rays, the same bright black-tipped central florets, the same wealth of beautiful dust blown about the hill-sides, clothing the honey harvesters also in the universal raiment of yellow, and realising Shelley's poetical star-fancy of "a swarm of golden bees." To be appreciated the life of the bush must be seen in association. A flock of cockatoos is nowhere viewed with such effect as against a background of red gums, with the glow of sunrise showing faintly through. The beating wings suggest a living snowstorm. From the river bends thousands of sparrows take flight together with a whirr of wings and a sounding rustle of thistle leaves like a hailstorm in dry reed-beds. With all his mischievous pertinacity one can hardly help admiring the sparrow. He is a type of the race—an energetic venturesome pioneer. Where other birds barely find a living he adapts himself to local circumstances and thrives. He has the old colonists' habit, too, of not caring much whose nest he seizes provided he is strong enough to hold possession against the original owner. Cornfield, vineyard, orchard, and thistle-brake are all his storehouse. He knows why the finest grapes are shrouded in white gauze, the largest cherries covered with cord netting. Perhaps the sparrow is never really out of place except when he masquerades at a city banquet as "lark in aspic jelly," and no doubt he feels it as much as we do.

The homing waters of the blackfish are generally along the outer curve of a bank. In these horse-shoe bends the dead wood lodges, the water-weeds grow, and the bank is more thickly clad with scrub. Since the forests have been cleared and the rains come to the sea with an unimpeded rush, the floods are made sharp and violent, and water-lilies have almost disappeared. Years ago the crystal pools were covered thickly in summer with floating oval leaves of light-green and white flowers, but now the water is seldom perfectly clear except in midsummer, and the water-queen rarely appears. The sharp points of land that from inner curves cut into the stream have outposts of thick rushes pushed as far as possible into the territory claimed by the other element, where they sway backward and forward in the current, always on the point of being overborne and drowned, yet always surviving the danger. Across this cape the wind drives sharply as the stream around its point, and just as there are no water plants where the current is strongest, so the scrub plant, the snow-bush, the Christmas tree, the acacia, and prickly box, cluster, bound together with twining acacia, in the sheltered bends, and rarely venture out to select a home on this wind-swept bank. The Christmas tree is in a sense the counterpart of the holly of the home countries. As the scarlet berry gives its ruddy colour to Christmas decorations in "the old country," so here the creamy blossoms of the Christmas tree are the only shrub flowers that survive the blaze of midsummer. The river light-wood is king of his kind. In Gipps Land his long, spare trunk is crowned with strange mosses and lichens which find a dwelling-place in his rough bark, and food in the perpetual damp of sun-deserted dales. Here the lightwood is rich in colour, robust in form, and in every leaf an acacia king. The clematis here is the small-leaved virgin's bower, with a larkspur-shaped blossom of the delicate tint of crude silk. Though its fragile tendrils cling lovably about companion shrubs, its embrace is selfish and fatal rather than affectionate. When the clematis annexes a bush that bush, sturdy as it is by comparison, must surely die. Yet its death gives new beauty to the forest, for the fratricidal clematis is, for the greater part of the year, a pillar of blossom, and a pleasant spot of colour in the gullies.

The number of plants along this river bank not native to Australia is somewhat surprising, but the near presence of certain long-established farms and artificial pastures no doubt accounts for their existence. Years ago the Indian castor-oil plant, with lilac trumpet-shaped flowers, and a prickly seed-case more formidable even than the husk of the chestnut, grew thickly along the banks, but now scarcely a plant is seen. All these strangers seem to come and go in cycles. They multiply too rapidly, and dying finally of their own exuberance, make room for something else. One year the banks are red with sorrel, next season the wild camomile struggles with it for possession, then the dock joins issue, and finally the spotted thistle, with its beautiful variegation of silver flake, overwhelms them all.

There are stray plants of fennel in the moist bends, but they are poor colonists, and make slow progress. Of course our own silver wattle, always beautiful as universal, lines the stream, and as the flame tree of New South Wales gives its ruddy blaze to northern landscapes, this tree marks out a continuous floral line for the river. Few trees have evoked more genuine admiration than the wattle in bloom, and few are more deserving of it. Outside the tropical Queensland forests, the scarlet flowering gum of Western Australia, and the Waratah, of Blue Mountains fame, are its only rivals.

If the river be low during a dry summer, the little native fish, locally known as trout, may be seen in shoals everywhere. In their pretty coat of white and green, they are quite unlike the home trout, but one rare variety of native fish, known as the spotted trout, resembles it more closely both in appearance and habits. Its form is very much that of the English trout, and its markings are similar, save in colour—black spots in relief against ashy brown. These fish lie under a shelving rock or in the hollow of a log, and rush out when an insect floats by or a bait is drawn past them through the water. When the yellow-flood waters were creeping up into bends and by-channels it was curious to note how the colours of the fish changed with the water. Several other fish beside the trout had their dwarfed prototype in the stream. There was an exact *fac-simile* in form of the ocean flathead. Then a beautiful little perch, with bright crimson fins, seldom grew much longer than the shrimps, with whom in the weed-beds they were found in company. There was also a Tom Thumb herring, with shining scales— a perfect copy of the sea-fish.

Of late, the many freshets in the stream I write of have prevented the growth of water-lilies and weed-plots, and the tiny fish scarcely survive. Many years ago strangers, supposed for a time to be herring, appeared for a season in the stream, but their peculiar smell when taken out of the water—that of a freshly-cut cucumber—betrayed them to experts as members of the English grayling family. The platypus makes its home in every reach and bend of this stream.

The other day one of the finest specimens of this strange creature I have ever seen was caught under the willows at night while trying to remove the bait from a fishing-line, and after living a fortnight fastened to a log, died for want of proper care. It fought hard before it was landed, grasping the water-weeds and snags with its webbed feet. It spent its period of captivity principally in burrowing, and with proper attention would certainly have lived. The spurs, like bits of polished amber, proved this platypus a veteran. It is said that the wounds made by these weapons are no more injurious than the spurs of a game fowl or spur-wing plover. It may be so, but there is a little tube running from the base close up to the point for whose *raison d'être* there seems to be no explanation if not intended to secrete some poisonous fluid. When the water is clear the platypus is rarely seen except about dusk. Then it comes from his burrow beneath the bank, and floats quietly about on the surface of the water. When there is a freshet in the stream it floats about all day.

Just opposite the willows there is a high red bank, along which hundreds of the little white-breasted fairy martins build their nests. When there is a heavy flood in the stream, the water flowing swiftly along the bank cuts its way into the soil, leaving the high-water mark clearly defined. Under this ledge, which forms a roof for their homes, the swallows build. On a warm day, and in the nesting season, when the clay dries quickly, the birds, flying backward and forward, mark out an almost continuous line from the bank to a point at one end of the willows where there is a low muddy piece of shore. Here the birds congregate in scores, kneading up their mortar, and carrying it away in little lumps to the bank. They make remarkable progress, and the day's work is always perceptible, through the mortar being still damp and darker than the other portions of the nest. The pear-shaped nests with the little circular entrance at the smaller end are sometimes built in a long line, and sometimes in a cluster.

The swallow is an intelligent little architect, and always shapes his nest to suit the position. Before the introduction of the sparrows no bird was so safe as the white-breasted swallow in his little castle. His position midway up the bank defied approach either from above or beneath. No native bird able to pass through that tiny doorway ever interfered with him; but the sparrow, a thorough vagrant by instinct, converts the comfortable nest to his own use, and is strong enough to hold the fort against the original owner. As the swallow is wise enough to choose the highest flood-mark for his building lot, it is only once in several years that the rising waters reach his home, and then at a time when the nest is not in use.

At the foot of the bank, amongst the rustling reeds, there is a shy warbler, often heard but rarely seen, whose eggs, resembling in size and colour those of the sparrow, are often addled by the flood-waters; but still the bird keeps to the same plot of reeds. The effect of its obstinacy is shown in the rarity of the species.

The wagtail builds a little cup-nest of spider's web

and grass woven together and fastened firmly on a bough. The sticky spider's web worked through the whole makes an excellent cement; the nests hang together for years after the birds have deserted them, and when four fully-fledged young birds, each almost as large as the nest itself, are packed together inside, it is well that the sides of the home should be able to bear some strain. A tumble would be fatal here, for the nests are generally built on a bough stretching over the water. When the spot is approached the parent birds show their alarm by constant cries and flutterings. If the intruder remain quiet for a moment, the female—the lighter of the two—after fluttering over and about the nest for a few minutes, settles upon it. Other birds are more suspicious. You never seem by any chance to catch one of the honey-eaters at home, although scores of them are twittering about up amongst the blossoms of the gum trees. The pendent nests, like wicker-work baskets, are so thin that, even a few yards off, they scarcely show amongst the branches.

Some years Willow Land is thronged with a species of migratory bush-martin. They come in large flights, and on a fine day their arrival is made known by a sharp, twittering chorus when the birds themselves are so high in air as to be barely visible. Their habits are those of birds that merely loiter by the way. A few twigs thrown together on a bough form the nest, the three large eggs are rapidly laid and just as rapidly hatched, and the young birds grow at a marvellous rate. Then one day both old and young disappear, and you see nothing of them, perhaps, for years. In the following year, as the seasons vary, they may nest farther north. With a score of the tiny birds that build in the bushes by the stream there is never any lack of music under the willows, and they sing as cheerfully on Sunday as during the rest of the week.

The willows grow as freely here as though to the manner born, and no one can tell from whence the first slip came. Those that line the Upper Yarra are said to have sprung from a twig brought by a former governor or military official from the willow growing over the grave of Napoleon at St. Helena. The weeds of one country are the flowers of another; the hyacinth that we treasure in our gardens covers the English meadows so thickly that Tennyson speaks of "sheets of hyacinths." And this same sweet-smelling hyacinth is the cammas flower that Joachim Miller tells us "covers the valleys of Oregon in spring-time as with a blue mantle." Where the banks are steep tufts of native geranium grow in the niches left by the roots of giant trees that flourished and faded centuries ago. The pink flowers blotched with crimson are sweet-smelling, but the soft, velvet leaves of the plant are even more fragrant. Above the bank, in the ridges of barley stretching between the orchard trees close down to the willows, there are tares and yellow mustard plants, and pink vetches to lighten up the bands of green as poppies do the English cornfields.

A Day in the Bush.

To those who first experience the freshness and rare elasticity of the country atmosphere in the early morning it comes as a revelation. If we care to imitate Nature, and reject the artificial habits that have grown upon us, a fine example is offered to us by every bird and animal in the bush. They all anticipate the sun, work through the dawn, and wisely sleep through the trying periods of a sultry summer's day.

The hare has spent the night in feasting on tender barley shoots in the fields near the river, and now saunters away up to the table-land, where its form is hidden by the long brown kangaroo grass. In the cultivated lands farm labourers will keep moving about during the day, and here pussy's rest would be disturbed, but up on the plains only the patter of an occasional sheep nibbling the neighbouring tussocks breaks her light sleep. The hare has a weakness for following a beaten track, and in that respect resembles slightly some especially orthodox divine. The difference is that in the one case the beaten track is supposed to lead one clear of all the snares that Satan has set for erring humanity, while in the other it leads straight into the snare set by the poaching schoolboy. Hares and rabbits, although resembling each other so closely, show two distinct and very clear types of animal development. Living at one time evidently under the same conditions, they both, when compelled to seek shelter from their enemies, choose natural depressions in the ground, and when these were not deep enough they were hollowed out to suit the proportions of its body. From this starting-point the two types seem to have diverged—the one going deeper into the ground, its organisation adapted itself to new conditions, while the hare, still trusting to flight, developed opposite characteristics. What can be more suggestive of speed than the lithe, graceful limbs of the hare, and its long body, offering so little resistance to the atmosphere? While, on the other hand, what can be more indicative of industry and ability than the muscular little fore legs of the rabbit when the animal is at work on one of its long tunnels?

Of all the forest noises that salute the morning none are so pleasant or melodious as the carol of the magpie. To speak by the book it should be called the "piping crow shrike," but few save technical naturalists are unkind enough to overburden a very sociable bird with such a pedantic name. The early colonists gave the name to the bird because it was like the magpie they used to see about the fields of the old country; if it wasn't exactly alike they couldn't help that, and a magpie it will remain in popular nomenclature until we become more learned and less indulgent. Up on the topmost spray of a gum tree, or with his black-and-white coat outshowing distinctly against the great yellow bloom-laden wattle boughs, he watches for the sun and breaks the vigil with occasional long-continued bursts of melody. With his breast thrown forward, head, back, and bill pointing aloft in disdain of everything earthly, he throws bursts of pipe music into the clear morning air, and the note being taken up by one bird after another, the solo finishes in a grand chorus. The bird is protected and respected by everyone except the Vandal schoolboy, who will shoot anything that moves, and has besides a grudge against the magpie on account of the sturdy way it defends its home in the nesting season.

A forest walk to be thoroughly enjoyable must be taken on a spring morning after a night of gentle showers. Then the woods seem to give out all their balsamic odours, and the atmosphere is full of life and vitality. The perfume of the forest flowers tones down the resinous exhalations of the towering trees, and intermittent zephyrs laden with the delicate odour of native musk, all combine to charm and elevate the senses. Many parts of the Otway forest, running parallel with the southern coast, are, as I have said before, almost untravelled. Their summits are as dense as when they first caught the eye of early Dutch navigators first ploughing the Southern Ocean. The wealth of vegetation is bewildering, and the novelty-hunter discovers rich stores. Down in some of the dark silent dells, where a gleam of sunlight has not entered for perhaps a century, there are strange vegetable forms. On the boles of some of the great tree ferns a dozen parasitical ferns of smaller growth are sometimes found. When the stream that now gurgles beneath the shadow of the water-plants was swollen by a winter storm into a torrent, the fern spores from the higher lands were borne down until caught by the rough trunk of one of the giants of their race. Here, undisturbed by any other agency for a season, they have sprung into life, and have established a miniature fernery on a fern. Some very beautiful native flowers are found amongst those timbered heights. Orchids, less striking than the gaudy butterfly imitations of a tropical forest, but always pretty and attractive, spring up in the bare spaces between the trees, looking like many-coloured stars in a sky of green. All the native flowers fade quickly. Once severed from the parent stem, no artificial means can prolong their life. They are born of the forest and of Nature, are an essential part of the wild system of the ranges, and repel civilising influences.

The well-meaning enthusiasts who offer prizes at horticultural shows for bouquets of wild flowers merely force unfair comparisons by placing a bouquet of wild flowers, from which the glory has all departed, alongside a collection of comparatively hardy and long-lived exotics. The beauty of native flowers can never be appreciated until we find them blooming like the splendid and striking desert pea amid the whiteness and desolation of a northern tropical plain, or giving a welcome relief to a monotony of green in the wooded wildernesses of the south.

The diffusion of plants is no new topic. Grant Allen and other English writers upon botany in its modern phases have long since laid down the general tenets of an advanced botanic creed. I only propose to mention a few of the peculiarities of Australian vegetable forms, together with some of the agencies that on this side of the globe assist in the spreading of plant life.

A great many of the wild plants of Australia are of the same order as the common thistle, the seeds being attached to the downy, inanimate wings, which without vibration convey them wherever the winds ordain. You may see them borne on the gentlest of zephyrs, the seed downward like the car of a balloon, and in the same way acting as ballast to secure evenness of flight for those long delicate wings. Anyone who has been in the country must have noticed how rapidly, and with what ugly results, the Scotch thistle has spread through the colony. As the name implies, it is not native to the soil, and the origin of all the vast thistle-fields is said to be a little packet of down brought here by a male Pandora in the guise of a patriotic Scotchman, to remind him of the national emblem and his own heather hills.

The Cape weed, another invader, also illustrates the remarkable efficacy of this special contrivance in Nature. In the spring-time some of the country lanes within a few miles of Melbourne are on a sunny day all ablaze with these golden flowers. When the crown of the bloom fades, the lower parts of the pistils change to a light velvety down, and on these soft wings the seeds are borne to adjoining fields. So effective is the method, that in favourable soils whole pastures are in a couple of seasons covered with the weeds.

The seeds of the familiar "milk thistle," some of the common marsh weeds, and the majority of wild, soft-wooded plants are furnished with the same appliance, and the more the winds blow during the seeding time the more complete and general is the distribution. Prevailing winds carry the seeds in a particular direction, and in this way many vegetable zones are accounted for, altogether apart from climate influences.

Very important distributing agents in Australia are the flocks of sheep which travel from one locality to another. This was especially noticeable along the main roads leading to the Melbourne markets before the railway system gave such complete and general means of transit as it does at present. Then large flocks of sheep travelled along every country road, and camped on every open pasture and common, and in these places strange vegetable forms were continually springing into existence. A like phenomenon was observed in France after the Franco-German war. The German cavalry brought strange varieties of seed in their forage sacks, and new plants sprang up where they had camped on French soil. In many cases in Australia, the natural conditions not being favourable, the strange plants have died away again just as suddenly.

Years ago, in some parts of the Keilor plains, I have seen solitary plants which I afterwards found were native to the greater plains of Riverina, and in some cases these lone pioneers have gained a sound footing in the new country. The seeds of many native grasses seemed to have been specially designed to assist such a method of distribution. Anyone who has walked over a plain in summer-time when the grasses have ripened, and are withering white under the scorching sun, must have noticed how easily the seeds, when ready for shedding, adhere to anything with which they come in contact. In taking almost any section of the wooded banks of the Deep Creek at least a score of such plants may be found. The casing of the seed is covered with small, intrusive barbs that penetrate any suitable substance quickly enough, but are not so easily withdrawn. After such a walk you will find attached to your clothes perhaps a dozen varieties of grass seed, thus illustrating in a small way the perfection of minor methods of seed distribution.

When the wild life of Australia was more common on these plains than it is now, the kangaroo and other furred animals all played their little part in perfecting the scheme. Now the domestic animals, hunting dogs included, which have supplanted and outnumbered them a hundred-fold, have unconsciously taken up the mission, and what could do better service in this way than the fleece of a merino or long-wooled sheep? The grass seeds penetrate it easily, and the roughly-coated seeds, such as the trefoil, and a number of burrs, all catch hold of it in their tenacious way, and keep possession for a long time.

In one plant especially the method of transferring the seeds is very perfect. The crown of the flower when faded resembles somewhat the down of a thistle, being even on top. Each separate hair is, however, barbed at the summit, and should anything brush against it the whole of the down, with the seeds outward, clings to the new object.

Berries such as the wild raspberry and others, that grow so thickly along the banks of most of our Australian streams, when not eaten by birds and so distributed by what is known as the best and most universal of Nature's methods, do not all fall upon the soil beneath, to give birth to a horde of young brambles that may engage in a struggle for existence with the parent plant. These brambles favour the banks of small creeks, and are seldom found in arid situations. When the fruit—or, to speak more correctly, the bunch, for there are a score or so of independent little berries, each with its own seed, composing these pretty scarlet fruits—wither, and are shaken from the stem by a gust of wind, they fall upon the waters beneath, and float miles away before they drift ashore, and in time give birth to a new raspberry bed. One day last summer, when strolling along the banks of the George's River, near Lorne, I came upon one of these wild raspberry thickets. Towards the head of the stream the brambles were stouter and more plentiful, not from any difference in the soil, but because they were probably much older than the others. Those up-stream were evidently the pioneers of the species in that particular spot, and the others farther down-stream were "their children and children's children." Sometimes these brambles, as well as many other wild plants, are found in depressions bordering on the river banks, or about the old channel of a stream that has slightly changed its course. The seeds have been deposited here by the waters at flood-time, when the rivers overflow their banks and intrude into every adjoining hollow or piece of low ground. A piece of low-lying land thus periodically flooded is certain to be covered with a thick scrub. Many writers on botany have commented on the fact that the generality of wild fruits are of such a size that they can be swallowed easily by birds without being first picked to pieces, thus allowing the seeds to fall to the ground uneaten. In the latter case Nature's object would be defeated, for while the birds would still eat the fruit, they would give no return by carrying away the seeds also.

When trees bear fruit so large that they cannot be swallowed easily they are generally called solitary trees, not because they are difficult to propagate, or do not bear freely, but because the inconvenient size of the seed is a bar to their distribution. The quandong, otherwise as edible and attractive to birds as most wild fruits, may be taken as an illustration. Had the "wild cherry" been nearly as large as the garden fruit of the same name (but not the same species), we would never have found such beautiful groves of these handsome trees as are to be seen between Dandenong and Ferntree Gully, as well as in many other parts of the colony. Had the stone which grows outside been at the reverse end of the fruit the birds might have eaten the latter without interfering with the seed, but actually the stone first presents itself, and it is so small as to make it a matter of no moment to the bird whether it swallows it or not.

Many human peculiarities seem to have their prototype in the plant world. In old folk-lore—the natural parent of the modern language of flowers—every well-known floral specimen represents some sentiment or other; but for the most part they are not chosen through any special fitness. There is no method in Ophelia's assortment when she says: "There's rosemary, that's for remembrance .... and there is pansies, that's for thoughts." Even the fatalist has his sluggish, suicidal creed represented. Many of the heavy seed-bearing annuals have no provision for distributing their kind. The plants live through a brief season of vigour and then perish, the seeds falling thickly around the white withered stem, and bursting into a nursery of tiny plants in the succeeding spring. Then, from their very infancy, the heirs of that careless, selfish plant engage with each other in a struggle for existence in the very harshest sense of the term, for to fall behind in the competition means death. In such a strait the human fatalist sums up the situation while charging a pistol, and the misguided plant commits suicide in another way. In their confined space the young plants grow sickly, but ere they can deteriorate so far as to be a reproach to the race the flowers refuse to fulfil their mission, the myriads of fertilising insects pass on to other fields, the germinating power is lost, and when the members of this unfortunate family die fighting each other to the last they leave no successors to continue the feud in the following spring.

One of the most remarkable branches of the plant world is the familiar mistletoe, that long ago became an institution in England at Christmas-time, but is rarely torn from the trees to shrivel under the burning suns that mark the festive season in Australia. What is a mistletoe? From where does it come? Some have considered it the out-cropping of a departed growth habit of the tree where it is now found; but the more general and popular theory is in favour of an independent, if parasitical, plant. A skilful gardener may by grafting cause entirely different fruits to grow upon the same tree, and Nature manages it without any such operation. The mistletoe offers to the birds that flit about the trees whereon it grows attractive berries, and thus achieves its object. The berry is covered with a stiff holding vegetable glue which enables it to adhere to bark and branch. It is only when thus exposed that the mistletoe will germinate at all. Planted in the soil the berry is quite sterile. The rough bark of the sheoak and its soft sappy wood offer a double advantage to the mistletoe, by giving the seed a point of attachment in the first instance and to the plant sustenance afterwards. In a sheoak forest, therefore, mistletoes are always plentiful, and after the sheoak they flourish most freely on red gums. Their roots penetrate farther into the sappy bark and soft wood than with most other native trees, and as more of the sap channels are thus intercepted and converted to the wants of an alien, the mistletoe growing on the sheoak is generally a vigorous and healthy plant. And it requires to be so, for the tree is very often found in exposed situations, where its peculiar foliage offers but a slight resistance to the winds, so that a gale tearing away the branches of other trees only plays a plaintive forest melody as it comes in contact with the long waving plumes of the sheoak. In return for the service of distribution carried out by the birds, the mistletoe gives them a protection for their homes, both against the elements and against their natural enemies, which the open foliage of the sheoak could not afford. In the middle of the tuft, amongst the thick fleshy leaves, the nest is sheltered and hidden, and in a sheoak forest almost every bunch of mistletoe is thus occupied.

When in some of the forest tracts of Victoria we see grand trees towering up from every loamy gully, and just monopolising sufficient space to bring them to maturity, we are apt to wonder how it was all brought about. It would seem as though, ages ago, some mysterious landscape gardener had allotted to every tree its area, and said, "Thus far shalt thou go, but no farther." If the soil was obliged to support double the number, they must necessarily be dwarfed in size, but every tree seems to have just sufficient space to bring it to perfection. How is it that in this wealth of vegetation no arrogant young usurpers spring up to contest the position? Simply because the older tree, being in possession, has pre-empted a certain section, and has in its own roots the remedy for aggression. The means for preventing encroachment upon their domain are the acids in the roots, which poison the soil in their neighbourhood for other trees, and thus convert it solely to their own use. In shallow country the rich soil is near the surface, and here the roots rush out rapidly on all sides, taking up a large area, so that the country is lightly timbered. In very poor land the trees take such selfish possession of the soil that not even a dwarfed shrub and scarcely a grass blade can live within the charmed circle. The selector in a timbered country, without troubling himself about cause and effect, is aware that if he destroys the tree the grass will grow, and therefore he "ring-barks" his timber. In deep rich country, where the roots strike far down into the earth, the older gums are more sociable, and sometimes allow a number of dwarfed trees to take possession of the surface soil in the shape of scrub, and struggle with each other for existence, while the old giant looks complacently down upon the contest.

One noticeable point in the forestry of Australia is the disappearance of the sheoak groves that once studded many of the southern plains. In the old days, when all Australia and a good percentage of the old world was flocking to those new Eldorados, the Ballarat and Sandhurst mines, it was a novel sight to see groups of kangaroo sitting solemn and stately in the thick shade of these trees, where the brown stone walls now intersect the Keilor plains. The wild game have entirely disappeared, and the aboriginal is only occasionally represented in these parts by either an utterly degraded and hopelessly vagabond member of the race, or one semi-civilised. The sheoaks also are a dying race. They are declining before civilisation. In the plantations and hedgerows we sometimes find young trees, but in the open pastures, where groves have been naturally established, they are all dwindling to decay and dying without heirs. They have made way for more aggressive usurpers in the vigorous Pinus insignis, or the silvery cedars that originally flourished on Lebanon and other historic mountains in the Cradle of the World.

Although a summer night in the woods is always enjoyable and beneficial, its glory palls beside that of an Australian winter's night—a clear, frosty, southern winter's night—a night when every planet adds a special brightness to the calm glory that is around and above us, and every emotional being is forced to the realisation of his own small individuality by the beautiful omnipotence of Nature. There is a glitter and sheen from every inanimate object under the magic moonlight, and the gross and abrupt in Nature are invested with a sober grace that accords well with the spirit of a dreaming reverie. As we stride along the grassy slopes a hundred dewy diamonds are whirled from their couches in the hollows of the leaflets, and flash round us in a jewelled shower surpassing in beauty the supremest creation of art, while about our moving shadows on the grass the moon throws a golden halo something like the glory that the old masters of art long ago cast round their ideal of the Saviour. On the crest of the ridge, loading the air with sweet incense, the great gums stand out in bold relief, a mass of grandeur, defined against the steel-blue, star-studded background.

The most popular of Australian authors has noted the "weird melancholy" characteristic of these forests in repose; and the realisation of it comes to us intensified by the voices of the night. The startling scream of the curlew on the uplands has a yearning ring in its rising cadences that invests a night-dream with old recollections. But the memories induced by that pathetic cry of the night bird vanish quickly as they came, and to increase the sense of utter loneliness the solemn call of a mopoke comes up from the deepest and darkest corner of the glade, where the dense foliage of the ti-tree makes a gloom, and the tall swamp plants shoot up so fragile and tender that the slightest touch brings them rustling down shattered and destroyed. We can imagine a note of despair in the sharply intoned call of this half-mythological frequenter of the night. His "never more" rings out persistently like the comfortless chorus of that "stately raven of the saintly days of yore," when it interjected its obdurate truisms into the fanciful pictures of the poet's meeting with his "lost Lenore."

Now another sound comes to our ears from afar through the clear air. A cry ike the shriller notes of a pipe, heard at regular intervals, and increasing in volume with each succeeding moment, but throughout invested with the same melancholy cadence that pervades every whisper of the night. It is a flight of wild swans winging their way to the south. For an instant, in the golden rays of the moon, we catch a glimpse of those flashing pinions, and then they pass away over the tree-tops and beyond our sight or hearing. Down the creek the dingoes are out on their nightly hunt, and an occasional long quavering howl, comprising a new scale of notes, for the expression of abject misery, pain, and despair, is borne with an accompaniment of echoes to our ears. This is the nocturne of the Australian bush.

Anon a shadow sweeps across the illumined sward, and we have a faint glimpse of a solemn owl borne on silent wings out beyond the limit of vision.

> "The eye
> May trace those sailing pirates of the night,
> Stooping with dusky forms to cleave the gloom,
> Scattering a momentary wake behind;
> A palpable and broken brightness shed,
> As with white wing they part the darksome air."

The Australian owl is a mute hunter, and his complaining hoot is never heard from "ivy-mantled tower," or hollow oak, or barn, or deepest shadow of the wood. Looking upward towards the myriad of radiant worlds with our own familiar constellation, glittering to-night with increased splendour, we have a better conception of the immensity of the universe than daylight affords, and may well feel in a humble mood— the reaction perhaps from that dense egotism which once prompted man to regard this grand system as a subsidiary illumination created for his personal convenience. But now the moon has set, the curlew has ceased its wail, and over the calm face of Nature, as an impenetrable veil, falls softly the darkness that cometh before the dawn.

Suburban Walks.

How many people are apt to think that, for the naturalist, suburban fields have long since been deserted, and that wild life can only be studied in remote places, far in the bush. I have spent days in the wildest mountains in Victoria without seeing as many birds as may be found in strolling through a field by the sea where the city clocks can almost be heard striking the notes. They were chiming a six-o'clock morning medley as I went out of the railway-yard a few days since in one of the suburban trains. A little more than a mile distant from a railway station there is a small creek trickling over a sand-bar into the sea. The tide comes up for half a mile, and at the flood there is sometimes a good fishing-ground here for a few hours. The banks on one side are matted with dwarfed ti-tree scrub, giving good shelter to the rabbits; and farther up, where the tide never comes, are quiet pools fringed with sedge and bulrushes.

Had the fish been feeding this morning I should

have given little attention to the surroundings; but shrimps have no attraction for the silvery bream, and a sand-worm cannot tempt the freckled mullet to destruction. A walk down the shore towards Laverton and back through the fields is, therefore, a more pleasant prospect.

There are many sea-birds about the shore here, and the rougher stretches of coast-line have also a special interest for conchologists. First a group of silver gulls are startled into flight, and then a white-breasted cormorant, lately swimming half under water, leaves for other haunts. A quarter of a mile up the beach one of the big black cormorants—a bird dark, lonely, and cheerless—is perched on a pinnacle of rock, stretching his neck at intervals, as these birds do when alarmed or suspicious. Only those who have dissected one of these birds killed after an hour's fishing can have any idea of the enormous quantity of fish it destroys. In flight one of the old birds keeps in the van, with the others stretching out behind him in a perfect wedge-shaped figure—a formation that cleaves the air more easily, and assists their flight.

Amongst sea-shore birds the distinctions between species are tolerably subtle. The tiny sand-piper that runs before the pedestrian on every bit of waste open beach, is the smallest of the order, and it merges into the dottrels. Science only knows how many dottrels there are. Even with those native to this bit of coast ornithologists have to resort to such faintly distinguishing terms as "hooded," "shaded," "banded," "little," "big," &c. Of course, when the name indicates some special trait in the habits or appearance of the bird it is difficult to find fault with it. But inexpressive terms, such as shaded and banded, big and little, have small significance. The largest of the dottrels is very like a snipe, and the snipe resembles closely a larger bird of the same order. In fact, the point where one class ends and the other begins is not always apparent. The difference between the dottrels and sea plovers is also slight.

Along the shore are flocks of a species of bird which some sportsmen and the gamesellers in the city are pleased to call snipe. They are probably tringa, a branch of the sea plover family. At sunset they feed about the mouth of the creek here, but find little peace now in haunts where they once prospered.

Many birds native to this bit of beach and meadow have been driven farther back to the great swamps on the plains, where, with a few swan and wild ducks, and perhaps a pair of blue cranes, they form quiet little communities. Some score of miles back along the course of this little creek, where it winds about amongst the stony mounds that break the level of the Keilor plains, I have come across the nest of the blue crane. The banks are lined with a few dwarfed gums, and the tree-building birds all over this wide tract go there in the nesting season. Every tree is occupied by a nest of some bird or other. In one tree a tuft of dry sticks with very little lining carries the light-green eggs of the crane In the next a pair of magpies hold absolute possession against all comers. Farther down two big brown hawks have appropriated to their own use a forsaken crow's nest.

Under the very shadow of this hawk's nest a pair of yellow-tailed wrens built one of their pendent nests. Evidently they had no fear of the hawks, who ignored such small game. Close by a branch was saddled with the mud-nest of a pair of grallinas, who, in a conspicuous coat of black-and-white, piped shrilly about the waterpools.

These latter birds have a cup-shaped nest, something like that of the house swallow when building on a ledge, but the plaster work on the outside is smooth, and being more exposed to the sun, dries very hard. The eggs are speckled like those of the minah, but are slightly larger.

Before leaving the coast for the meadow let us see what shells are to be found along the beach. Like the shore birds, they have a very close alliance. Gathering up a specimen of every variety to be found upon the sand and the rocks, an ordinary observer would probably range them under five or six heads, with slight variations in each; but the scientist extends the classes to perhaps fifty. The limpets, although they seem to be one family, come under many classes in the collection. It is in the study of the smaller shells, however, that we realise how universal is the life of the sea. In the bottom of a little pool left by the flood-tide there are hundreds of tiny shells, some so small that their shape can only be decided with the microscope. Take up a handful and you have a perfect mystery of species. To the naked eye they are almost all alike, but under the glass distinguishing marks come out. These tiny shells are in themselves an extensive study. Of course there are mussels along this shore, and under that title we have a huge family, with representatives on every coast in the world. In South America exists a perfect marine giant, yet in appearance it is a fac-simile of the little black mussel lying here on the sand. The unio, another branch of the family, with a white pearly cup on either shell, is found on the sandbanks of inland creeks as well as here on the sea beach. The sea specimens are a longer oval and more irregular in shape, and the lines on the shell curve less gracefully than the fresh-water specimens. The handsomest shell to be found here, however, is certainly the phasianella. Its shape is that of the variety of garden snail curling upward instead of inward to a point. The markings on the outside, which is smooth and bright, are very beautiful. Indeed, the shell has an outer coating like enamel, and in this its beauty lies. The shading in the background varies from the darkest brown to a light buff, and two shells are rarely found with the same colour and markings. There are stray specimens washed up upon the sands here, but their favourite bit of beach is on the inner side of Swan Island. Near the little torpedo pier at the north end of the island they abound, and those who spend "the season" at Queenscliff and wish to turn their attention to shell-gathering will always find something there to repay a journey. Those shells, and indeed salt-water specimens generally, should never be washed in fresh water, lest they lose their bright colours. It is the same with the little scraps of scarlet, and white, and pink seaweed tossed up by the waves. In making up a specimen-book these should be pressed as soon as they are taken from the sea water, for in that way only are their brilliant colours retained. Another shell that clings hard to the rocks here is the white turbo, with very handsome orange shading on the inner side. This shell-fish has a hard brown shield, fitting closely into the aperture of the shell, and protecting it on its only weak side whenever it has been displaced from its hold on the weed-covered sandstone. The phasianella's shield is pure white, bright, and hard as ivory. Where the creek trickles over the sand there are hundreds of semi-transparent, fragile shells, in colour a dull white, speckled with pink, but the name of which I have forgotten. Of the trochoidæ, another beautiful class of shells, I find a single specimen—a conical shell that seems to be made up of one reddish-brown thread winding up to a point. Here also is the common English whelk, rare in Australian seas. The smaller specimens are cream-coloured, and serrated on the outside like the murex. Another specimen is the "bishop's mitre." The mitre may have been originally modelled on the pattern of the shell; the shell was certainly never copied from the mitre.

Amongst the refuse of the sea waters who can fix the limit and say, "Here the sea animal ceases and the seaweed begins?" Now that nerve centres have been found for the sponges another point has been gained by those who would extend the "animal" kingdom indefinitely. Amongst the specimens to be picked up here, none are more remarkable than the sea urchin. You find only a fragment of the animal, barely sufficient to give a suggestion of its real appearance. The empty shell, generally called a sea egg, had an animal inside, and outside a forest of spines for its protection. It had a mouth out of all proportion to its size, and possibly an appetite to match. Its teeth were large enough to hollow out the little pits in the submerged rocks where it lies. In other waters the sea urchins are found close in shore, their spines bristling above the edge of their pits like the bayonets of entrenched riflemen. There, under the leeside of rocks, sheltered from the rush of the ocean rollers, they live their own quiet circumscribed lives. Some scientists say that in certain stages of their development—for, like the butterfly, their youth is a time of transition—they closely resemble the asteria or starfish.

Having followed the beach far enough, we may strike out amongst the sheoak trees and come back through the fields. The influence of the cold ocean breezes is noticeable here. Winter lingered long by the shore before he was driven back south to his own white land, his ice caves, and snow mountains. Then spring came tardily down from the tropics, and so her flowers, dead long ago in "the interior," are yet beaming here in the grass. The first flower to catch the eye is the very nearest approach amongst Australian wild flowers to the English dandelion. And once more one may recall the rapid spread of English plant-life in this southern world. Hundreds of wild weeds, which originally found their way here in seed shipments, may be seen in the Melbourne Gardens, while those with winged seeds have spread far into the country.

Waiting the other day at a suburban railway station, I counted between the metals nearly a score of foreign plants and grasses. There were clovers (yellow, white, and oval-leafed), sorrel, dandelion, thistle, docks (the "soor dockens" of the North country), Cape weed, lucerne, a variety of milk thistle, horehound, mallow, a scrap of English rye, and a tuft of American prairie grass. All these aliens, scorched every minute with the grimy breath of throbbing engines, were flourishing, with scarcely a native plant or blade of native grass amongst them. It was typical of Anglo-Saxon colonisation.

The first sign of field life is an insect—a butterfly, with bright auburn wings starred with black and set off with a sombre edging. There are not many butterflies hereabouts, and the naturalist who wishes to study them must take his net to the mountains. Our Victorian butterfly is much plainer than those up nearer the tropics; and the moth, although larger, is plainer too. Yonder is a hawk perched on the fence. By its long legs, and what can be seen of its shape in the distance, I take it to be the handsome goshawk with large yellow eyes and beautifully barred breast. He wears handsomer plumage than any other member of the tribe. Farther up on the plains two of the smallest of colonial hawks are very common. The kestrel haunts the thistle beds for mice, while the grey kite with its sparkling ruby eyes skims along the grass. The kite has a white breast with slate-grey on the back, and if not so well proportioned as the kestrel, has very powerful wings, and makes even more havoc amongst small ground game. Larks, always found in the vicinity of towns and villages, are plentiful in the grass. The scientist has again trenched upon the liberty of the subject by calling them pipits, but they are larks in plumage and in habit, while at certain seasons of the year they sing and soar as sweetly as any of the race. The bird has the backward spur, characteristic of all the lark family, and if anything else is wanting to fix its identity, place it in a cage alongside the British lark and see both birds prove the brotherhood, as captive larks do, by running to and fro and fluttering against the bars in a way that must excite pity in the breast of anyone but a bird-fancier. Years ago when I was a boy in the country one of our neighbours caught a lark and placed it in a cage fastened to the verandah. All day we watched it with its breast against the wicker-work beating the cage with its wings. That night someone guided a fishing-rod to quietly push out the peg that fastened the cage door, and in the morning the lark was with its fellows on the hill-side welcoming the sun. Skylarks singing in the early summer when the shadows of the white clouds chase each other across the waving corn—is there anything sweeter in field or forest? The skylarks disappeared from our Victorian plains for years, but they have come back again, and upon any fine day you may hear them sing.

In a little grassy bend where the creek winds before running over the sands a single plover is in possession. I call it a plover from force of habit; and a plover it will always be in the country, just as the stone plover is a curlew—but the peewit is its correct name. For every bird of the kind now out in the grass country there was once a thousand, but they have been killed or driven away. They learned to suspect and avoid man, but, like the wild turkey, were never suspicious of a horse, so they were shot in scores from the saddle. The young plovers run about as soon as they are hatched, and the little downy chickens in the grass look like tufts of rich brown and white velvet. With anyone close by they crouch quite motionless, and but for the too evident anxiety of the old birds one would never suspect their whereabouts. It is not easy to find them in the grass, and if the eye is diverted for an instant they are probably never seen again. Young quail and hares have the same faculty for hiding themselves in the barest pastures. The plover builds no nest, but lays its eggs—in colour an æsthetic green with brown spots in relief—in any slight depression. In a small patch of white rustling reeds some distance up the creek a pair of the reed warblers that make music for anglers along the Saltwater River have built their home. Their eggs, placed in a big cup-like nest, resemble very closely those of the common sparrow. Amongst the furze bushes at one corner of the field there are some tiny blue wrens. In the nesting season you find their home in a little warm ball of moss down at the foot of a bush. Out in the sheoaks little, red-billed, red-tailed finches build a very large nest, but with a circular passage so long and narrow leading into it that nothing larger than the owners can enter. The little grass paraquet fluttering up from amongst the flowers is not showy enough to excite admiration, so it is left unmolested to hatch its white eggs in the crevice of a sheoak or dwarfed gum tree, while its more brilliant cousins are caught and caged.

Coming back through the meadow, I find a little bloodsucker. What evil fancy prompted such a hideous name for the tiny little fellow hunting for insects in the grass? Certainly, if you point a stick at him he opens his little yellow mouth to its widest limit in the most defiant way, but this is a threat only, and his solitary act of aggression. On the coarse bark of the sheoak you sometimes find a larger one with a tail twice the length of its body. It will bear the very closest inspection without moving, and seems to depend for safety entirely to the fact that the tree trunk it has chosen for a hunting-ground is exactly the colour of its own rough skin. These inoffensive little creatures bear this nightmare of a name which even serious science has adopted.

Melbourne has, undesignedly, been built on a natural borderland. To westward the plains, on the margin of which we have strolled, extend away for many miles. Now let us look eastward, to where the timber thickens gradually into the density of a Gipps Land forest. The gums have been rudely replaced by garden trees, and the wattles by willows. Just now in September the shrubs in many a wide-spreading domain help materially to give colour to the prospect which is daily brightening.

"Oh ! gaily sings the bird, and the wattle boughs are stirred
And rustled by the scented breath of Spring."

A mild winter is not conducive to a perfect spring. Flowers are half born when the late rains and spasms of cold, the more bitter through being long withheld, pinch the tender buds and they perish. The result is a blighted, intermittent spring. When, however, the winter has not been a "broken" one, the buds burst, with one impulse, into flower, making perhaps a briefer but certainly a better spring. The acacias are amongst the first to herald the coming of the season of flowers, and the wattles of northern forests that have been acclimatised round about Melbourne so far cling to old traditions that they bloom much earlier than the black or the golden wattle. Frazer's wattle, a Queensland acacia with an iridescent tinge in its foliage very like the leaves of the saltbush, is about the first to put its suggestion of gay yellow into the green sobriety of a southern winter garden. Only the Murray wattle, with its great, woolly, fragrant blossoms, is more beautiful. A New South Wales acacia, carrying its bloom in spikes instead of the broad racemes of the golden wattle, runs its Queensland cousin a close race. The banksia blooms linger on rather as surviving flowers of winter than harbingers of spring. In the cold season, when other blossoms are dead, these are the salvation of the honey-eating birds. People who live in the south of Victoria know how the wattle birds congregate by the sea when the honeysuckles are in blossom. Along the eastern shore of Port Phillip Bay, in the very heart of winter, one is amazed and delighted with the variety in species and plumage of the honey-eaters. On occasional sunny days they fill the groves with the life of their tireless wings, the sparkle of bright feathers, and the variety of a note not always musical. In the Botanic Gardens the honeysuckle bed has always life and colour at this season of the year. One of the finest of the scented shrubs and trees grouped here is the sand heath from the mounds above Botany Bay. Late in the summer I have found it blooming beautifully upon the heathy hills thereabouts. It thrives wonderfully in our climate, and just now is at its best.

The blossoms that jut up so freely between the hard, sharp, burnished leaves are tinted in richest shades of red and yellow, and full of a sweetness both of nectar and perfume. Where the breeze curls round the tree into an eddy the air is suffused with a subtle fragrance that tempts one to linger and inhale. There are quite a score of New Holland honey-eaters wrangling musically with each other for the right to dip into the choicest blossoms. Their black bills, long and slender, and slightly curled downward, seem incapable of the constant clatter that marks the warfare amongst the flowers. Shafts of yellow run along the line of the quill feathers with the wing extended, and cluster into a distinctive patch of colour on the side when the bird is at rest. On the breast this honey-eater is pencilled like the wattle bird in white and black. Fewer in numbers are the "white-eyes" that dart quickly from bush to bush close over the grass. The editor of a lady's newspaper would call the colour of their coat an æsthetic green; a bushman would understand it better if described as mallee green. Other occupants of the honeysuckle grove, lighter in colour and larger in size than the "white-eyes," are the common white-plumed honey-eaters, which flit in pairs, while the "white-eyes," though they cluster together on the same bush, dart away singly and at intervals.

The bird's-beak Hakea, of West Australian forests, is conspicuous from the red flowers growing so strangely, not as terminals to the younger branchlets, but clinging to the bare limbs. Another species of honeysuckle, the protea of the Cape of Good Hope, is dotted with late blossoms, and into their thorn-girt chalice—its very appearance suggesting rich stores of nectar—the honey-eater dips freely. The most unique of the group is the Indian Hakea. Every flower rests upon a leaf which curls up lovingly about it in the form of a green hood, enfolding and protecting the blossom.

Flowers of September, in another sense, while hardly heralds of spring, are the popular ornamental foliage plants. What pillars of colour in a winter garden, for example, are the ruddy Japanese cedars. Riding through the county of Evelyn a few days since, I noticed a profusion of yellow in a sheltered hollow. Viewed more closely they were merely the last leaves of the sweet briers, yellow and ready to fall, yet still clinging to the stem and lingering on through the winter months. There was more than one migrant brier in this hollow. Along one side were clumps of hawthorn growing more freely than I ever saw them elsewhere, and covered with clusters of reddish purple fruit that gave a predominant hue to the colouring of the bank. Farther down were tufts of furze—not the native furze, which belongs to the honeysuckle tribe, but the English yellow-flowered thorn plant. How did they come here? The question was answered later on. Within half a mile I found a couple of cultivated paddocks bordered with hedges of hawthorn and sweet brier. Amongst the haws the bright plumage of some rosella parrots was conspicuous. They flew before me along the line of the hedge, and finally reaching the end went screaming away to the hollow. Here was the explanation of this little British colony in one of the gorges of Diamond Creek. These parrots were the messengers that had carried the seeds to the hollow, and unconsciously planted for themselves a winter orchard. The little white seeds of the sweet brier are distributed in the same way, and the hedges, if uncared for, spread out, taking absolute possession of the grass lands. The furze is independent of the birds, and more rapid in its outward march. When the seed-pods burst in the heat of midsummer the seeds are cast far away on either side. All day you hear the mimic fusilade along the hedgerows, and see the shining little seed-bullets shot out into the paddock and the roadway.

Among flowers designed only to give colour to garden and landscape, the September mosses fairly claim consideration. In long-settled districts where the rail fences are falling to decay they are always plentiful. Out about Bundoora they cling so thickly to the grey posts and panels that the old wood is almost hidden in this soft wrapping. Where the rails face the west they become mere lines of lichens, in which many tints of green and much variety of form appears. Down towards the earth the posts on one side are tinged red and yellow with hundreds of tiny wood fungi. With respect to mosses, each locality has just now an individuality. Down about Laverton, where a couple of tiny streams trickle across the plains to the sea, a dwarfed ti-tree, clinging low about the ground, like the gunyang or kangaroo apple, borders the banks. In the gaps of the scrub the rocks are thickly covered with the most brilliant of lichens—in colour a burnished orange, and gleaming brightly amongst the general dulness of green. Perhaps the sea-salts, in both air and earth, contribute to this brightness. Take another locality—Beechworth, the city of bay trees. In its gullies the yellow boulders—

piled in the lavish style which Nature adopts for her monuments of stone in the gold country of the north-east—have an envelope of pea-green moss. Here it also gave a distinctive colour to the gorges, which are bold and sterile below, though the tops of the hills are capped with rich red soil and pleasant vineyards. Out about Keilor plains the lichens of the fences are a sort of red mushroom with the cap inverted. The basalt rocks, their faces grey with the weather stains of centuries, are covered with moss, the varieties of which develop gradually into dwarfed cliff ferns. You find these tiny ferns clinging where the moisture trickles down through a fracture in the wall of rock. Here they have the seclusion that most of the ferns love, while they get also a filtered share of the surface food. The main ground of the common moss is a sward of plush velvet. At intervals little rush spears spring, which terminate in an oval fruit, yellow at first, but reddening as it ripens.

The departure, rather than the coming, of birds indicates for city people the change of seasons. In the ponds by the Yarra the wild duck and coots are yet plentiful, the former as they float motionless seeming a mere tuft of dull brown feathers on the water. The Australian black duck or teal at rest is a typical object, its head being lost in the general outline of the body. It cannot be confused with the garrulous domestic duck fossicking so industriously amongst the water-reeds. When the wild ducks leave the ponds spring and their nesting season is close at hand. They go to the quieter bush swamps and streams to breed, and return to town in the Christmas season, when the fowler's gun once more wakes up the echoes of their country home. In these city ponds the black swan builds amongst the white reeds, and one of them may now be seen upon its nest. The coots, tame as they are, object to anything like prying curiosity. You may pass close to them without exciting much notice, but a steady stare discomposes the bird. His four-note call becomes more and more frequent, the white tail so jauntily tip-tilted flips more industriously, a light trumpet-call, something like that of the black swan in flight, breaks in upon the regularity of the complaint like the false note of a singer, and finally the coot takes flight for a tuft of weeds in the centre of the pool, screaming in time to the flap of his wings, while his long legs dangle awkwardly about the water. One of the reed warblers keeping company with the porphyry coot has a peculiar note like the slow, opening rattle of a mower at work in the hay-fields—that measured, mechanical click heard while the horses are being wheeled slowly from one stretch to another. Imported birds seem to mate rather earlier than our own songsters. A few days since I watched a pair of starlings surveying the corner of the Melbourne Museum in quest of a home amongst the ivy. One of them, seated on the mason-work of the building, chirped out a warning that some too curious person was beneath, and his mate at once came from his retreat, and the pair flew away across the grounds. In ten minutes they were back again for a further survey. Here, as in other lands, the swallow is a harbinger of spring. When the hundreds of fairy martins, the smallest of all the swallows cluster about the high red banks of the river, the foundations of those mud nests that are so much larger than the bird itself will soon follow. As the birds flutter about the bank they keep up a constant twittering chorus. Listen to a number of ladies in a room whispering in groups, while some person, not a man, is playing the finest of sonatas, and if you can stand it you have a perfect imitation of the fairy martins and their lisping silvery whistle.

A pair of swallows are always associated in my mind with no less serious a matter than the break-up of a church. This church was a selector's hut of the kind that both squatter and selector build but seldom inhabit. Hither an harmonium had been brought from one of the homesteads in a waggonette, and in these journeys it had also been bumped hopelessly out of tune. A sheet of iron on the roof was always loose, and it flapped ceaselessly in the wind, keeping measured time to the drone of the hymn. I remember above the pulpit a pair of welcome swallows sat by their nest, the most attentive of the congregation. It was my duty to describe that church for a certain newspaper, and wishing to convey an idea of something primitive in architecture, standing alone in the pastoral wilderness, I called it a "dummy hut." Perhaps, in my unconscious "new chum" ignorance, I had gone very near the truth—the truth which both pastor and people felt, but which never found expression. The sketch did it. Never again could the parson—a real archdeacon—pray or preach in that hut, never again the people listen. Next Sunday they went out under the wide-spreading arms of the trees—a cathedral more appropriate. Though no censers were swung in these box-gum aisles, the scents of the woods were sweet. The swallows kept possession of the old hut while the disjointed roof played its old tune to their little ones.

He who travels much by rail will find in the winter night work of the selector—the annual " burning off"—something that speaks also of preparation for spring. Rushing over the monotonous level of the wheat country, one sees through the windows, and far away, the thousand lights of a phantom city where no city is. The illusion is almost perfect, the lights being, perhaps, a bit too red for gas jets, except when seen through a fog. There is nothing so deceptive at night as a bush fire.

One night I stood, with some fifty other residents of a country village, and watched for hours the mountains burning, thirty miles away as we calculated. How magnificent we thought it—that is to say how magnificent it would have been had we been closer. We made every allowance for the distance, and everything thus wanting in effect imagination supplied. One of us had seen Mount Disappointment crowned with flames, another had witnessed the burning of the Sebastopol plateau, and so experiences were multiplied.

Suddenly, as we gazed and wondered, a strange phenomenon was wrought. A giant form loomed up behind the fire, and with one stroke swept away miles of mountain and burning trees. Whereupon the saddening truth struck us. The short, dry grass on the opposite hill-top, not more than half a mile off, had been burning feebly, and someone passing by was brushing it out with a bough.

We came down from that hill silent and pensive. Nature and the night had played a practical joke with us, and each of the victims felt ashamed of himself, and angry with his neighbour.

Something about Snakes.

WHILE popular delusions are being every day destroyed the snake has thus far escaped critical examination. Like that perturbed spirit whose business is to disturb the peace of mind of orthodox humanity, it has gained a world-wide notoriety while really doing very little to deserve it. Armed by Nature with terrible powers for retaliation, the snake has used them with the greatest discretion, and yet man, partly through fear, but largely on account of a fervid imagination, has given it a pre-eminent place amongst the foes of man.

Snakes will bite, and at least fifty per cent. of them fatally, when you force them to it, but as a rule when they come in close contact with man they are just as eager to get out of the way as he is, and that may be taken as a high compliment to their activity. There are odd members of the human race whose curiosity would prompt them to look down the barrel of a gun in order to study the early stages of an explosion. These are the people who in the bush go clawing after imaginary rabbits and hares in a hollow log or a chance burrow in the bank and discover through being bitten that it is the country residence of a snake. The snake of fiction is an unsightly, venomous thing that sits in impossible attitudes on a small section of its tail and hisses malevolent defiance at the whole world.

Now and again, if bush stories are to be believed, it casts off this martial air and descends to such undignified domestic occupations as milking a cow, helping itself at table in the hearty hospitable back-blocks fashion, or standing tip-tail in the cradle for the edification of the baby. I can, however, hardly imagine a snake foolish or hungry enough to ruin its digestion by eating station "damper."

One might explain the value of the snake to the country journalist in the dull season, but I will not do that.

The snake of real life is so essentially non-aggressive that one is almost ashamed to say anything severe about it. In any country township you may find scores of men who have had unlimited narrow escapes, and who work themselves into cold shuddering fits in reflecting over "what might have been"; but those who have felt the fangs of a serpent—otherwise than in imagination—are few and far between.

He who has much experience with Australian snakes knows that they never strike "with intent," unless menaced by danger, or suddenly alarmed. I have trodden unconsciously within easy distance of many a coiled snake, and, in common with many more, have even stepped safely over them when hidden in the grass; but only in a single instance has a snake struck at me with murderous purpose.

One sunny day in summer, rambling through

a little rushy valley—an Eden, only lacking an Eve, and into which no serpent should have penetrated—a pleasant reverie was abruptly broken by a sharp hiss, the only time in all my experience that I ever heard a snake give such a warning. There was only a fragment of a second between that alarm and my spasmodic backward leap; but I had a glimpse of a venomous head and several inches of bared neck bending round in a rapid, semicircular sweep that just brushed my knee, and missed by the most uncomfortable of small spaces.

It was a six-foot brown snake, which had been lying out in a little break in the rushes enjoying the noonday heat, and perhaps absorbed in some little idyll of his own. I knew that I must have trodden upon it; but what with the nervous shock and rapid heart-pulsation of the moment, it was difficult to say what happened. That snake, like the average domestic servant, "gave warning," or it would never have been mortified by a miss, or have paid for the blunder with its own life.

In the same way I have only once seen a snake spring at an assailant. That was on a spring day in the Kyneton district. We were admiring a herd of cattle standing in the shadow of the trees, and completing as pretty a farm picture as an artist could desire, when we found that a full-grown brown snake was of the party. They are never scarce in those woods. One of us walked up close to it more carelessly than ordinary caution should permit, and raised his stick to strike, when the snake, without an instant's warning, sprang straight at his face. The style in which that man quickened up to his work would have been ludicrous had the possibilities been less serious. The leap of the snake was not more sudden than the fall of the stick, and the two met half-way with the usual result. The experience was enough to show that in attacking a snake it is not wise to make any hostile demonstration. Old bushmen, as a rule, carry the stick on their shoulder, and move up as if about to pass the snake on one side without noticing it. While he does this the snake's head is bent low amongst the grass as a hare hides its form, and if the man turns sharply, and strikes surely, there is no danger.

The average settler will make as good progress as most men in running for the doctor when a friend or relative is suddenly stricken with illness or accident, but if there is anything in the world that tempts him to linger by the way it is the opportunity of killing a snake. The very fact that those who come upon a snake in the country think only of killing it in the safest and most expeditious way is no doubt one reason why so little is known of their habits. But for their great fecundity the different destroying agents would soon have an appreciable effect on the serpent tribe, and we might as in New Zealand forests plunge recklessly through brake and bush with no fear of any coiled foe. The few proofs I have had of their rapid increase serve to show how many enemies are required to keep their numbers reduced to comfortable limits.

One evening my terrier "Tichborne"—whose conduct to the race has been already mentioned—"set" a snake most unmistakably. It was in a clump of those wild giant nettles that one lets go so hurriedly after clutching hold of them for assistance in climbing a bank. I only reached the spot in time to catch that distinctive rustling sound which betrays the movements of a snake of any size. It evidently had its underground home in the corner, and had taken shelter in alarm. A few days later, coming by the same track, I approached more cautiously, and this time was not disappointed. In the centre of the clump the very finest specimen of a tiger snake was coiled, for tiger-snakes run more to width than the others, and a five-feet snake is a large one. Just as it was about to glide away once more its head was blown to pieces by a charge of shot, for I had got so close that the muzzle was within a few feet of the reptile. Practically no less than forty-two snakes were destroyed by that single charge, for had the big tiger-snake lived a few days longer she would have been the mother of an interesting family of forty-one. The young ones were about nine inches in length, but showed only the faintest signs of vitality. A few days before the owner of the farm had killed another tiger-snake which had its home in a pile of vine cuttings. This one, under dissection with the pruning knife, revealed a family of twenty-eight, each about twelve inches in length. After lying for a time in a box exposed to the sunlight they became wonderfully active for infants, and glided smartly about their confined cage. On being pressed with a straw or twig they turned and darted their heads feebly towards the point of irritation. The instinct to bite was strongly developed, but the power which it controlled was yet dormant. They were an interesting family, but not sufficiently so to be granted their liberty, and like the rest of the tribe they were doomed to death. When alarmed and about to strike the tiger-snake flattens its head so as to bring its sheathed venomous teeth into prominence, and its appearance thus angered or alarmed is thoroughly savage. It was amusing to note how valiantly these infant snakes puffed themselves out in the same way.

The black snake was rare in that locality, but sometimes we found one—always conspicuous by his brilliant under-colouring.

A local cricketer was bitten one day while fielding on the playground in the centre of the village. The offending snake was only three parts grown, and the village brand of whisky applied as an antidote was more than a match for the poison.

. A good many of the pretty little grass-snakes are no doubt perfectly harmless. In summer they live like the lizards, upon insects which they find in the grass, and at the approach of winter work their way down a few inches below the surface of the ground, and lie there torpid while the cold weather lasts. On a hot day you catch sight of their silvery skins glistening in the dry grass, when yet too far off to be noticeable on account of their size. On a section of the Keilor plains they were at one time very plentiful, but the land was cultivated one season for a hay crop, and the snakes were almost exterminated. The owner of the paddock ploughed it towards the close of autumn, and it rarely happened that the ploughman opened a furrow from end to end without leaving behind him a tiny silver-skinned snake wriggling feebly in the sod. The exposure had the effect of killing them effectually, and when the land once more lay fallow there were no pretty snakes gliding about amongst the tufts of dwarfed kangaroo grass. In summer the hay-fields, and more especially those that run right down to the river bank, are a favourite hunting-ground of the snakes. In walking along the margin of these oat-fields, just as the corn is breaking into ear, and when the lower leaves have withered and fallen from the stalk, so that there is no dampness beneath, it is not an uncommon thing to see a snake disturbed from its siesta on the bare path glide in amongst the corn blades, where for the time it is safer even than in its own burrow in the river bank. In the mowing scores of them are cut to pieces by the machine. The small animals and birds such as quail and landrails move in gradually towards the centre of the square then being cut down, and if any of them come to grief it is usually in the last band of standing corn, where, confused by the rattle, they rush in upon the knives, and are mutilated. The snake, accustomed to the incessant clatter of the machine moving round about it all through the summer day, is only alarmed when the knives with their deafening din are close upon it. In the last moment it raises its head defiantly to strike this persistent intruder, and that movement is its ruin. In the next instant it is caught in the advancing teeth of the machine, and head and neck are lopped from the body.

I first saw a snake fossicking for mice one Sunday morning in a field where the hay-makers had finished their work and nothing remained but an expanse of shining stubble. On my first noticing the creature —a brown one some five feet in length—its head was hidden in the ground. I was soon aware from its movements that the reptile was invading the home of some other animal rather than seeking shelter in its own. It made several forward darts, as though trying to pin something in the hole, and then slowly withdrawing its head appeared with an unfortunate little mouse struggling in its mouth. The snake was no sooner clear of the hole than it became aware of some added feature in the landscape, and dropping the mouse, promptly turned its cunning little head on one side, as all reptiles do under such circumstances, and with its glittering black eye took in the situation. The interesting part of the performance was over, however, and a few blows from a pliant bramble brought down the curtain.

Towards the close of harvest, when the big stacks are rising slowly into the air all through the hay-growing country, hundreds of snakes again come to grief. Still following their game, the field-mice, they take shelter under the haycocks, and when these are finally raised by the field-elevator and the remnants of hay raked together to be thrown into the next dray coming for its load, the mice go leaping away amongst the stubble, but the snake, who has made the place his temporary home, is instantly killed by the haymakers.

Few living things are more easily maimed than a snake. The continuous vertebræ are very fragile, and the slightest blow means a fracture. If the injury is near the centre of the body the reptile is powerless to glide off, and can only lie there angry, venomous, and in pain to die by inches. A snake thus wounded is seen in its most savage mood. It darts its head now here, now there, in search of something upon which to wreak its vengeance, its little eyes fairly blazing with anger, and the black tongue darting rapidly in and out between the slightly-opened jaws. If a stick or any object moves within range it strikes viciously—the rattle of its teeth upon the wood telling how virulent would be the effect of a bite if any animate being were so unfortunate as to come within reach at the moment.

A wounded snake seems to derive a certain amount of satisfaction in thus venting his anger, for after striking two or three times he becomes either sulky or satisfied, and declines to expend any more of his poison. A snake injured in this way, and conscious that it cannot move out of the way, is really more dangerous than one with its faculties unimpaired, because of its greater readiness to bite at anything passing within reach.

In our old schooldays one of the favourite summer pastimes was to dig or root out a snake that had been "marked down" some days before; and when the doomed reptile was finally forced from beneath the rocks or dug out of the bank of earth in which it had made its home what a storm of missiles rained upon it from the ring of keen snake-hunting schoolboys. Of course many of the stones rebounded and struck those for whom they were not intended, but in the excitement of the chase these wounds were philosophically borne.

The great kingfisher or laughing jackass has on very slight evidence gained the credit of being one of the greatest enemies of snakes to be found in our forests. An observant Englishman who visited the colonies a few years ago wrote in a book, "The laughing jackass or great kingfisher makes night hideous by its insane laughter; in the day-time, however, it performs a very useful service in waging perpetual war against snakes." This is a very handsome compliment, but unfortunately not more reliable than his reference to a companion night bird. "Another bird that makes a great noise at night is the goatsucker, which continually cries, 'More pork! more pork!'" On many a still night in the bush I have listened to the weird metallic call of this strange bird, the mopoke of the natives, without hearing it give expression to anything resembling the pork-shop sentiment here attributed. If English writers would strictly report that they "had been told" certain strange things which they put into print about Australia we could understand it. Some bushmen have a wonderful faculty for invention when there is an intelligent stranger at hand to profit by it. I cannot myself admit that the hoot of the laughing jackass is either hideous or insane, nor, on the other hand, can it claim special credit for the destruction of reptiles. Out about Pastoria, in the Kyneton district, where both snakes and great kingfishers were plentiful enough, I never saw them in combat, nor had those who spent the greater part of their lives in the bush been so favoured.

On almost any hot day in summer the chattering of a flock of minahs away amongst the gum trees told that they had found a snake. The birds gather and keep up an incessant din, darting down close upon the snake with flapping wings and snapping beaks, and keeping the reptile constantly occupied. And while the sounds, which must have been familiar to every bird in the bush, were ringing through the forest the laughing jackasses sat stolidly in the branches overhead, taking no apparent interest in the proceedings. It is said that the jackass has been seen to carry large snakes into the air, and then allow them to fall to the ground, thus killing them. One reason for doubting the statement is that the laughing jackass has barely enough wing power to support its own weight, its wing feathers being short and soft, and the beat noiseless. Anything like a full-grown snake, whether of the brown, black, tiger, carpet, or diamond species, would at least be as heavy as the bird itself. A black hawk, with its long powerful pinions, might perform such a feat, but hardly a laughing jackass. That the laughing jackass destroys many small snakes may be admitted, but so do most Australian birds.

At one time we kept a young magpie in a cage after it had been taken from the nest on one of the gum trees in the garden. This prisoner was fed by the old birds long after its fellow nestlings had been compelled to make their own way in the world, and amongst other things dropped into the cage were occasional small snakes, some of them the little grass varieties, others young venomous snakes, which the birds must have killed down by the river-side and carried thither. The snake habit of frequenting the banks of creeks and rivers is fatal to numbers of them. After a freshet in the Deep Creek it was no uncommon thing to find dozens of drowned snakes lying amongst the rubbish at high-water mark. Just as the snake enjoys undue notoriety for ferocity, so amongst some credulous people it has gained credit for musical tastes. Meeting snakes of all kinds casually in the country, in my credulous days, I have at some facial inconvenience whistled various popular airs, without in a single instance causing them to assume the far-away look that marks a human being in the throes of musical enjoyment. Indeed, to be candid about the result of the experiment, the snakes invariably acted as though they had an appointment in some other part of the colony. Friends who wished to be thought critical have remarked that this was the very best evidence that snakes do really possess a sound musical taste. Perhaps they were right, for on that theory the absence of snakes from any hut where the farm hands have an asthmatic concertina, or from those farmhouses where an old-fashioned piano is kept, might be explained.

The serious drawback to an excursion through the heath country bordering the southern sea-coast is the knowledge that the trail of the serpent is over this Eden. A stroll through a corn-field, where, instead of wheat, there is a blaze of crimson, pink, and white flowers, is pleasant to the senses; but the romance evaporates when we know that the heath mounds are equally popular with the snake tribe as a promenade. Indeed, there are few parts of Australia where the habits of the snakes can be better studied than in certain sections of the southern coast-line, more especially the low land about Cape Patterson and Anderson's Inlet. Here the small streams from the mountains spread over the flats, and form a series of miniature lakes, where in the summer season the frogs sing all day in chorus, and fill the air with what to their uncultivated taste is, perhaps, music of exceeding sweetness. They usually commence with a modest solo, then a duet, a trio, a quartette, and swell finally into a grand chorus. In the very midst of the glee there is a sudden silence, and a score of splashes in the water tell that the choristers—with the exception of one who is struggling in the fangs of a snake, and filling the air with his terrified, croaking complaint—have sought safety amongst the roots of the water-weeds at the bottom of the pool.

On a warm day operations may be closely watched without disturbing either the hunter or the game. The snake is generally guided to the spot by the chorus of his victims, or a prior knowledge of the locality; and in taking up your station it is not advisable to concentrate your attention solely on the pool. One may have chosen as a look-out a point directly on the serpentine line of march, and in this case is apt to be startled at any moment by a rustling amongst the grass at his feet. These unrehearsed movements make matters uncomfortable, both for actors and audience, and break the continuity of the drama.

A position up amongst the branches of one of the sweet-smelling banksia trees that stud the coast-line forms an excellent look-out, for it gives a good bird's-eye view of the pool, and as the tree-snakes of Australia have their location farther north, there is nothing to flurry one's nerves. Sometimes the snake is the first to appear at the edge of the pool, and he gathers himself up amongst the marsh-button at the water's edge, not in the customary basking coil, but doubled in a continued letter S form, so that he can extend himself rapidly and with ease. Here he lies quite limp, and apparently asleep. Presently the green head of a frog makes a ripple on the surface of the water, and, after a short survey of the surroundings, strikes out for the shore. Fortunate is it for him if in paddling to the edge he comes in contact with the broad floating leaf of a water-lily. On this normal raft he may pour out his soul in song without any fear of foreign aggression. But the leaves of the lilies are not plentiful, so the frog generally gains the bank without interruption, and at once turns his back upon the source of danger.

If the frog would but sing with his face to landward, and practice the accomplishment of diving in backward, he would be all right; but his desire to take up a position from which he can dive into the water at the first alarm helps the hunter. If the frog lands close to the snake the latter loses no time in making the capture, but if he is out of reach the snake waits patiently until his game, with a few unwieldy hops, comes within striking distance, or until the strains of the first comer have beguiled another chorister from the depths of the pond. It is not difficult to fix the exact moment when the snake considers his prey within range. The graceful, bent, venomous head, which has hitherto lain flat, is raised, the bright black bead-like eyes shining, and the neck gracefully curved. The muscular sides quiver with a sort of suppressed energy, and the whole of the reptile's body, although maintaining the same position, is yet in motion. Then there is a sudden forward dart, a pitiful cry from the frog, and the head of the snake is raised still higher, with its prey struggling unavailingly for the curved teeth rarely loose their hold when once secured. Sometimes when snakes are plentiful and frogs scarce there is a fight for the game at the edge of the pool, a large snake often relieving a smaller one of its capture.

In *Chambers's Journal* there is a story of an Australian black snake chasing a clergyman at full speed across three paddocks. This black snake had the faculty of bringing head and tail together when in motion, instead of gliding along in the usual way—thus making a bicycle wheel of himself, so to speak. Those familiar with Australian snakes and with the effects of bush whisky will say either that this snake was no Ophidian, but the great original serpent, or that the reverend gentleman had evolved this conception of a wheel snake from the depths of his own boots.

Along the South Coast.

THERE is a sameness about the ornithology of the sea-coast of Victoria, and this is noticeable in the colour rather than the form of the birds. The white-breasted sea eagle has many of the characteristics of the humbler fisher birds that surround him, and he is, I should say, about the lowest type of his own brigand race. Amongst other birds seen along this coast is the sooty albatross, a smaller edition of the big, high-flavoured fellow, whose chief mission on the waters is seemingly to be trapped in the simplest way by the passengers of ocean-going ships. The different varieties of the family fill the same place amongst sea-birds that the swallows do upon land, and in shape they are veritable sea swallows. The crested penguin is a very remarkable bird. Nature seems to have given it a plume of no value whatever, while at the same time so curtailing its mere apologies for wings as to make these organs of no practical use. One might have expected to find in the man-of-war bird a gull with a stately "three-decker" aspect, but he is a long, low, clean swimmer, built on the lines of the modern torpedo boat. The big Pacific gull is no stranger here, and sometimes, well out in the straits, one has a glimpse of that lone *voyageur* the great petrel. Occasionally, after a continuance of southerly gales, fairy penguins are washed in upon the rocks, telling, as the dead birds did to Columbus, stories of other lands out beyond the green horizon. Among sea-birds the gradations from one species to another are more clearly marked perhaps than in any other order. The tippet grebe, besides having variations in its own family, is a distinct link between the cormorants and the penguins. As the conditions under which the sea-birds live are almost exactly similar, we find very little variety in the markings of their plumage. The prevailing colours are first of all white, then black, blue and grey, with a few nondescript shades as a result of the commingling of some or all of these colours. The predominance of white in so many sea-birds is strong evidence in support of the theory of that reign of universal white away back in the glacial period.

In Polar regions of to-day the conditions are

probably the same which affected our plains and mountains innumerable ages ago when the earth's surface was being changed by genial warmth from a cold white mass into a world of life, and colour, and beauty. And in these Arctic lands, more especially in winter, we find every animal and bird clothed in a garb of purest white. Possibly, the leading types of bird and animal life all the world over were similarly conditioned at one time, and the sea-birds, even those now frequenting our semi-tropical coasts, have retained the "glacial period" colours longest, partly because their natural habits have not undergone the same modification as those of most land birds; and more especially, perhaps, because their colour did not in any way affect their safety. The breasts of sea-birds are as a rule, white, but in swimming, that part of the bird is immersed in the water, and if not entirely so, it at least harmonises very much with the foam and spray in the midst of which the birds live. Even on land the colours agree quite as well as any other with the white cliffs upon which the sea-birds lay their eggs. The occurrence of albinoes, or purely white animals and birds, seems also to point back to a period when there was no variety in Nature. Near Corowa, in New South Wales, I have seen a pure white kangaroo, and fishing one day at the junction of two of the branches of the Deep Creek near Keilor was much interested in a real white sparrow flying about amongst the thistles, in the midst of a flock of its ordinary brown companions. There is no other bird of the same colour in Victoria which would be mistaken for a sparrow. These casuals must have been the result of an hereditary out-cropping, a reminiscence of the period popularly known as "Antediluvian times." Need I say that the white kangaroo was soon killed, possibly the white sparrow also? The hawks could hardly miss so distinctive a mark. Most breeders of pure stock have at one time or other been surprised to find a white calf in their herds, although the ancestors of the little animal for generations back wore a coat of fashionable roan. The colour may be a family trait, linking the little prize shorthorn of to-day with the white cattle that once ran wild through English forests. In the same way the albino connects the present with a yet remoter time. In both flora and fauna these peculiarities are noticeable. Those variegations which the modern gardener has marked upon almost every green leaf of the garden are but the intelligent cultivation of chance developments, such as those just noticed. In the sheltered valleys between these southern hills you will find dark damp nooks where the ferns and marsh weeds are white and tender through being deprived of a sufficiency of light and warmth. Place a young pot plant in a darkened room, and after a time the absence of those influences that have brought the plant to its present glorious perfection cause this little thing to revert to its hereditary condition of white feebleness. Exposed to the sunshine, it soon becomes green and vigorous again, and you have the metamorphosis that in the infant past was slowly wrought through countless ages.

Starfish are very plentiful along this rocky coast, and by raising any piece of flat sandstone below the high tide level you may find more specimens than you care to collect. If you wish to be quite correct in speaking of these little animals call them *Asteria*, and as the scientific name is, for a wonder, rather a pretty one, it may be commendable to keep to it. *Asteriæ* are more common, perhaps, than any other living forms in the world, for you find them on every coast. One of the offshoots of the tribe multiplies its arms in a rather remarkable manner, very much after the same fashion as the antlers of a stag multiply their branches, except that in the case of the starfish the multiplication seems to cease only with the animal's death. Commencing originally with five arms like other starfish, a second pair afterwards shoot out from the extremities. These again multiply, each successive output being smaller and more fragile than the last, until finally the arms become as fine as thread, and resemble a mere circlet of fringe. The multiplication, if ten times repeated, would give the starfish 5,120 arms. The variety just mentioned is native to the Indian seas, but with careful search you may pick up a specimen at almost any point along our southern coast. A friend was fortunate enough recently to obtain a very fine specimen alive, which he placed in his aquarium, where it was one of the chief attractions of a collection of Victorian curiosities. Shortly afterwards, while dredging in Port Phillip Bay, he brought up a beautiful harp shell, out of which, as soon as it fell into the boat, rolled a hermit crab, of a brilliant scarlet colour, and with a pair of fine eyes, blazing like sapphires. As a fine specimen of this remarkable order of *Crustacea* the crab was preserved alive and transferred, together with his harp shell, home to the aquarium. Here, however, he proved a veritable tyrant, for on the following morning the starfish was found with all its beautiful arms chopped into short lengths, like firewood piled at a forest railway siding. The hermit crab had evidently been on the war-path during the night, and the destruction of his tank-mate must have been a piece of wanton mischief, for there was scarcely anything edible about it, and, besides, the wants of the crab in the matter of food had received liberal attention. Since that time he has shown a disposition to destroy all and sundry, and some smart-looking fish, which in the water are much too clever to be caught by a sluggish hermit, are occasionally found dead and partly devoured in the aquarium. The nocturnal pirate steals quietly upon his victims during the darkness, and so effects their capture. Along the coast near Melbourne the shell collector finds little to reward his labour. It has not always been so barren, however, for in many of the cliffs about there are shell deposits in which we find the Spondylus, and other rare varieties. These are, however, the relics of the past, and no specimens of the kind ever roll in upon the sands now with the southerly breakers.

Sea aquariums are daily formed and daily absorbed by the falling and rising tide on every bit of rocky coast-line, and very beautiful are those situated just where the tide turns to creep up again over the water-worn rocks. The sides of the pool are handsomely decked with tresses of pink and white seaweed, adornments that can only exist here amongst the rocks. At Lorne one may spend many pleasant hours in seeking out these crystal pools, and studying the sea life that fairly teems in every drop of salt water along this stern Otway coast. In winter a great many marine animals seek the deep sea for warmth, but during the hot days of an Australian summer they come up almost to high-tide mark, and invest with a new interest the sea aquariums. The rambler amongst the rocks coming suddenly upon one of these pools has at first but a faint conception of the amount of life they contain. A few tiny fishes dart for the shelter of the seaweeds, or a family of young crabs scuttle off with their ludicrous side step to some hole in the rock. Beyond this, the pool seems deserted, but settling down for a time and quietly surveying this interesting tank one animate object after another breaks upon the vision, until the predominant feeling is one of wonder at our own first blindness. But, clearly, Nature never intended that he who runs should read here, for the colouring of the life of this pool is in such perfect harmony with everything else it contains that time is required to identify the living things. If it were otherwise, these coast pools would be rare feeding grounds for the cormorants and gulls. Perhaps the brilliant scarlet anemones are the exception to the rule, but one generally finds them attached where the shadow of a ledge falls, not so much for their own protection as that their presence may not be too apparent to their prey inhabiting the same water.

The anemones of this coast show, as they do all the world over, a wonderful variety in shade and colour. The scarlet kind, with its fringe of tentacles waving about in the water, is perhaps the handsomest of all.

Another, brown in the cup with a crown of green fringe, shows a preference for corners and retired nooks of the rocks. Others, again, are all brown and all beautiful. These anemones have a reputation as gourmands, low as they are in the scale of existence. One may prove the powers of the green-fringed variety by a simple experiment.

The rocks here at low tide are covered with limpet shells. If a number of these are detached, the limpets cut out of the shell, and brought on the point of a penknife one at a time in contact with these waving fringes—that seem so much more the attributes of a plant than an animal—the supple tentacles will grasp the morsels round, and the limpets slowly disappear in the pliant cup. Again and again I have seen the fringe extended, calling as plainly as possible for more, until fourteen limpets, about twice the original bulk of the feasting fish itself, had been swallowed. Then the waiter brought on a fresh course—mussels *au naturel*—and at the end of an hour six of these had gone the way of the limpets. Finally, having only a short holiday at Lorne, there was no opportunity of testing the appetite of this interesting animal to its limits.

The anatomy of the sea anemone is simple in the extreme. It eats, breathes, digests, and performs a half-score other functions with its stomach alone. And the sea anemone has—oh, enviable distinction!—no liver. "Picture it, think of it, bilious man."

While the anemone was being fed one or two scraps of shell fish sank, slowly whirling to the bottom of the pool, when suddenly a shrimp darted out from a cleft in the rock, and seized one of the morsels. With his long tender claws and feelers all spread, and the amber of his almost transparent body in faint contrast with the water, he was certainly one of the most interesting of the living objects of the pool.

It is said that the shrimp sometimes robs the slow anemone of his food, but this one showed no dishonest inclinations. Once, as he glided gracefully about the pool, he brushed ever so slightly against these waving tentacles, and seemed fairly electrified with fear at the contact. With a quick backward rush he was clear of the danger long before the closing fringe could envelop him. Perhaps the sharp spines of the shrimp were a protection to him; but not much in this case, I fancy, for the lower we go in the descending grade of existence the more faint are the perceptions of pleasure and pain, and the anemone is not very sensitive.

If by some accident his crown of fringe is shred away he grows a new one, or if he be cut in twain by a shifting rock, it means merely the creation of two anemones where one formerly existed.

The starfish has one distinction—he is perhaps the slowest mover of all sea animals, while yet being more nomadic in his habits than the anemone. One of the limpet shells, with fragments of fish attached, fell within a couple of inches of one of these reposeful asteria, and after an hour's toilsome travelling he got possession of it.

Some of the shell fish in motion about the pool, carrying their dwellings with them like the garden-snail, can, by comparison with the starfish, claim the travellers' right of being citizens of the world.

To westward of that headland at Lorne where the little pier runs out in the teeth of the seas there is a level bed of rock studded with flat boulders, the majority shaped like the loaves of bread baked at a country farmhouse. Under these boulders are myriads of strange sea things gifted with the capacity for living either on land or beneath the water, and upon this plateau, drained at low tide, the finest specimens of starfish are found. At the popular and growing aquarium in Melbourne it has been found that the starfish live only for a few days in the artificial pools. Could they be preserved a very fine collection might be transferred from here to the institution.

Under the same rocks with the starfish are huge sea snails, as large almost as one's clenched hand and with sides black as ebonite. The back is protected by an armour of hard white shell, and the colour underneath—a bright buff—contrasts finely with the shining black of the animal's sides. Another inhabitant of the rocks, slightly allied to the snail in habits and appearance, has a jointed shell which enables him to "double up" like the porcupine ant-eater of the bush, leaving no vital part exposed.

Another large and very interesting zoophyte of the rocks about this coast is the Cynthia. The best specimens are about the size and shape of an ordinary teacup, and rest upon a soft semi-elastic pedestal, the material of which closely resembles prepared india-rubber. The top covering is softer than the sides, and through this the animal—which completely fills the inside of the cup, and is of a dull-red colour—evidently obtains its food by suction. The inside of the cup is of a beautiful mauve colour, and this, with other ornamental peculiarities, make it a trophy eagerly sought by the summer visitors to the seaside. Those attached to the rocks at low tide are covered with a handsome fringe of seaweed, while others on the higher ledges are girt about with white corallines. How they got the popular name of Cynthia shells one can only conjecture. That beautiful sea-goddess Aphrodite might have used it as her drinking cup, and certainly could have chosen nothing more appropriate, but one fails to find the connection between these curious homes and the mythological huntress of Mount Cynthus. Many of the coast fishes found in the pools may be studied nearer home in the glass tanks of the Aquarium, but now and again one meets a rare specimen. One of these, a chubby fish, with his dark skin spotted with yellow eyes, not unlike the markings on a peacock's tail, would have shown to much advantage in captivity and gaslight. Amongst all the sea creatures that excite our admiration few claim a larger share than the daring little rockfish —the stormy petrel of the under sea. He is no lover of quiet pools and a tame existence. Where the foam is whitest and the billows lashed to wildest confusion by the rocks, there in the watery turmoil is the little rockfish. Those who seek for him must submit to being drenched by the spray, and flooded occasionally with a higher sea, but in spite of these disadvantages the rockfish gives, next to the bream, the best sport that the angler can get along this southern coast.

One of the recognised pastimes at this southern watering-place is the snaring of crayfish and giant crabs. Wherever there is a stretch of bold coast-line you are certain to find the different varieties of crustacea, the crab and crayfish apparently living together in harmony, except when there is a prospect of booty.

It is amusing to watch the proceedings of the two animals in such cases. As the crayfish is the great attraction from the gourmand's point of view, and the crab comparatively useless, the probability is that you wish to snare the former only. If at low tide you come upon a large sea-pool several feet in depth, and with the white sand disappearing beneath an overhanging ledge of rock, you are certain to find the crustacea there. From a weedy crevice, too, you may see their pliant spines protruding. Having made your bait, possibly a small bream or rockfish, fast to a piece of string, and weighted it with lead, you allow it to sink gently down until it rests upon the sands. If the pool is inhabited, ere long a crab or crayfish shoots out from one of the crevices in the rock and seizes the morsel. Sometimes the crayfish is scarcely in possession before a crab sidles up in his peculiar all-round style, and his supremacy being admitted without question, at once deprives his fellow-lodger of the prize. If the passage leading to their common home is a narrow one he drives the crayfish into it, and then taking the fish in his claws, tucks himself into the hole so as to block it, and in the most philosophical way proceeds with his banquet. His friends the crayfish are as effectually imprisoned as if a door had been closed upon them, and with those long claws of his as a standing menace they are not likely to play any mean tricks upon the master of the situation behind his back. The depth of sea-water is deceptive, and the novice in snaring crayfish is apt at first to make ludicrous mistakes. He insinuates his copper-wire snare gently but surely over the tail of the crayfish, makes a desperate stroke, and finds that he has been noosing salt water some two feet above his intended victim. Experienced fishers let the snare down until it cuts a tiny groove in the sand.

The sponges enjoy a negative distinction, as some scientists have allotted them the honour of being the very last possibility in the way of organic bodies. And although it seems much too early to make a definite boundary-line between the organic and inorganic world, the sponges certainly look the part allotted to them. Anything more expressionless, or more abjectly helpless, cannot be imagined amongst the things that have a being either by land or sea. A flower that offers a gorgeous couch to the passing insect, and then closes up and destroys him as soon as he takes possession of it, might reasonably be granted the power of sensitiveness, if not a capacity for calculation : but why the sponge should enjoy the same distinction is not exactly clear. Some authority, more gifted than others, has now said emphatically that the sponge is an animal, but qualifies the bold assertion with the opinion that it is a very inferior sort of animal indeed. While the scientists differ on the point, we recklessly hold our own opinions as to its being a sea-mushroom, a water-lichen, or something of that sort. Anyhow, the sponges of Australia are amongst the prettiest and most interesting of the treasures of the deep. They are parasitical in their habits, and establish themselves on any submerged object, so that sometimes we find them in novel positions, either enveloping a shell or themselves caught and bound in threads of coral. You may make a fair collection of sponges in shallow water, or stranded on the beach after a heavy gale, but the best of them are found many fathoms deep, and are only to be got by dredging. The Australian seas are especially rich in sponges; and some rare varieties already known to science with many entirely new specimens have been found in the waters of Port Phillip alone. Strange to say, there is nothing like a complete collection in our national museum, although Mr. Bracebridge Wilson, of Geelong, who has given a great deal of attention to sponges, as well as other branches of marine life, has sent many cases of them to the British Museum, and delighted some of the specialists at home with his new contributions. Probably, in the next work published on the subject, we shall learn something of the products of our sea-waters. In order to get an idea of the real beauty of the sponges, they must be cut with the dredge-knife from the rocks on which they grow far beneath the surface, and plunged in a living state into the methylated spirit necessary to preserve them. If required for ornamental purposes only, you may pick them up in hundreds along this coast-line after a stiff gale from the south, but if the opportunity is long delayed those washed ashore are covered with the shifting sea-sands and lost for ever.

At certain points along the shore one notices great mounds of limpet shells mixed with ashes. Sometimes they are covered with drift sand, and appear to be nothing more than ordinary coast dunes, but when the surface is removed shells and ashes are found piled in confusion together. These signs indicate that at one time the camp-fire of a beach tribe of aborigines smouldered here for a period, and so the tons of partly calcined limpet shells were brought together. Some of these mounds are worthy the attention of an Australian antiquarian. Even the most cursory search reveals, for example, the pointed bones which the aborigines used as forks for the roasted shell fish. These are formed of the smaller bones of the kangaroo—the large bone of the hind leg the natives used to force the clinging limpets from their hold on the rocks.

The coast is the geologist's favourite field of research, especially where the outline is bold and rocky, and where the breakers have cut quaint figures in the sandstone and revealed a notable variety in strata. Along these cliffs the diligent searcher for fossil remains is most likely to meet with something to reward his labours. Perhaps it is a piece of petrified wood, the fragment of an extinct vegetable world. In yonder block of sandstone it has lain hermetically sealed until the intrusive persistent waves once more brought it into the light of day. What changes have taken place between that interment and resurrection! In some parts where the cliff has been undermined and broken away through the action of a mountain stream cutting its passage to the sea, we find the most beautiful impressions of ferns and seaweed of a variety altogether unknown to botanists. One almost fancies that such apparent perfection of outline and reproduction of delicate fibre is due to the existence of organic vegetable matter, which remained invisible until the action of water developed an outline, even as certain chemicals or agents, such as heat, etc., change a blank sheet of note-paper to a highly-interesting missive when chemical ink has been used.

Where the fissures in the cliffs have been coated

with a few feet of soil, a growth of rugged scrub comes into existence, and in these sheltered places we generally notice a hardy little coast wallaby dart for shelter as soon as any strange object appears above the ridge. He has not always lived amongst the tussocks of long wavy grass and the general vegetable dreariness of these exposed and precipitous cliffs, but has been compelled to seek the spot as an asylum. The marsupials of Australia are regarded by naturalists as strikingly illustrative of the theory of development from a common source. It is a long chain from the big forester, down through the different varieties of wallaby to the kangaroo rat, and finally, to the tiny interesting little creature known on the plains as the "kangaroo mouse;" but all have the same characteristics. Back beyond high-water mark, where the sand has not encroached, the mounds are all covered with beautiful heaths—the epacris of the botanist. In the brilliancy of a spring day the bells expand, and from the main lines of crimson and white the colours converge towards each other through all the beautiful shades of dark and pale pink. These purely Australian flowers claim a near kinship with the ericas, sometimes seen to such advantage in a cottage garden; but amongst the many British and South African varieties of this hardy shrub, all brought to a high stage of perfection by the gardener's art, few surpass in beauty our own Australian heaths in their natural homes.

Printed by Cassell & Company, Limited, La Belle Sauvage, London, E.C.

www.ingramcontent.com/pod-product-compliance
Lightning Source LLC
Chambersburg PA
CBHW021359230426
43666CB00006B/573